Fibrous Polymeric Composites

Fibrous Polymeric Composites

Environmental Degradation and Damage

Bankim Chandra Ray

Rajesh Kumar Prusty

Dinesh Kumar Rathore

CRC Press
Taylor & Francis Group
Boca Raton London New York

CRC Press is an imprint of the
Taylor & Francis Group, an **informa** business

CRC Press
Taylor & Francis Group
6000 Broken Sound Parkway NW, Suite 300
Boca Raton, FL 33487-2742

First issued in paperback 2020

ISBN-13: 978-0-367-57157-3 (pbk)
ISBN-13: 978-1-4987-8401-6 (hbk)

Library of Congress Cataloging-in-Publication Data

Names: Ray, Bankim Chandra, author. | Prusty, Rajesh Kumar, author. | Rathore, Dinesh Kumar, author.
Title: Fibrous polymeric composites : environmental degradation and damage / Bankim Chandra Ray, Rajesh Kumar Prusty, Dinesh Kumar Rathore.
Description: Boca Raton : Taylor & Francis, 2018. | Includes bibliographical references and index.
Identifiers: LCCN 2018000458 | ISBN 9781498784016 (hardback : acid-free paper) | ISBN 9780429506314 (ebook)
Subjects: LCSH: Fiber-reinforced plastics--Deterioration.
Classification: LCC TA455.P55 R385 2018 | DDC 620.1/92323--dc23
LC record available at https://lccn.loc.gov/2018000458

Visit the Taylor & Francis Web site at
http://www.taylorandfrancis.com

and the CRC Press Web site at
http://www.crcpress.com

Contents

Preface

Recently, fiber-reinforced polymer (FRP) composite materials are being used for many components starting from casual household items to critical engineering components and structures. The extensive acceptance of these materials comes from their structural tailorability, light weight, and cost-effective manufacturing technologies. Advancements in these materials have significantly contributed toward their potential exploitation in high-performance and high-precision mobile and immobile structural applications including aerospace, automobile, marine, sports equipment, bridges, buildings, and so on. The most attractive strength of composite materials lies in their superior specific properties (e.g., strength-to-weight ratio and modulus-to-weight ratio) in conjunction with good impact strength, corrosion resistance, fatigue durability, and damping characteristics, which motivates engineers to use these materials in a wide spectrum of diversified applications. As the application spectrum of these materials is quite wide, the interaction of these materials with their application environment is quite decisive on the long-term durability and reliability. Ageing and in-service environments may be static or dynamic in nature. Ageing due to different environmental conditions is essentially fading away their durability and reliability. The environmental parameters may include temperature, moisture, ultraviolet light, and other high-energy radiation (electromagnetic, microwave, γ-ray, etc.). These parameters play important roles in altering the physicochemical structure of the polymeric material and, hence, the integrity and durability of the components in its in-service condition. In this book, an attempt has been made to illustrate the durability of these FRP composites at various sets of environmental parameters. The first chapter has been mostly focused on the critical applications of FRP composites and the nature of harsh and hostile environments under which the materials are exposed. Chapter 2 deals with some of the characterization techniques that are essential for experimental determination of the extent of the interaction of the materials with various environments. Presence of the polymeric phase, as well as interfaces, limits the elevated and high temperature, as well as low and cryogenic temperature applications of polymer matrix composites. In this context, Chapter 3 is devoted to assessing the interaction of environmental temperature with the FRP composites. Environmental moisture acts as a plasticizer for most of the polymeric materials and, hence, deteriorates their mechanical strength and stability. Various physical and chemical interactions take place between the water and epoxy molecules. Interaction of environmental moisture with polymer and polymer composites and its subsequent effects on the characteristics of the composites have been discussed in Chapter 4. In some cases, there may be a fluctuating environmental temperature during moisture absorption, which may create a thermal fatigue or even may cause freezing of the absorbed moisture. The effects of such dynamic environmental parameters on the performance of FRP composites have been illustrated in Chapter 5. Further, in several critical applications, the FRP composite materials may get exposed to high-energy radiations like ultraviolet rays, γ rays, and other electromagnetic waves. Similarly, in various space applications (low earth orbit), there is a continuous exposure to atomic oxygen, micrometeoroids, electromagnetic waves, charged particles, and other man-made debris. These harsh environments may pose a threat on the long-term durability and integrity of polymeric composites. The effects of some of these environmental parameters on the structural durability of FRP composites have been discussed in Chapter 6. As with conventional metallic structures, FRP composite materials are also sensitive to strain rate. The strain rate sensitivity of these materials is discussed in Chapter 7. The addition of the carbon nanotube kind of nanofillers to modify the polymer matrix of the FRP has been broadly acknowledged to enhance various thermal, electrical, and mechanical behaviors (especially out-of-plane properties) of the bulk material. The effects of the addition of carbon nanotubes on the environmental durability of CNT-embedded polymer composites has been explained in Chapter 8. Finally, Chapter 9 is focused on some of the characterization techniques to assess the damage tolerance of FRP composites and various possible means for its improvement.

The authors have tried to provide some glimpses of the interaction of the FRP composites with various types of possible environments. This book is evolving to provide a platform to the researchers and engineers who are working on FRP composite materials.

The authors would like to take this opportunity to extend their heartfelt gratitude to the National Institute of Technology, Rourkela, India and its bright and beautiful-minded pupils. These pupils conceived and compiled a few words on the environmental damage and degradation of the advanced structural FRP composites and illuminated the current cutting edge research on the nanofiller additions to it. We would also like to thank the generous support and suggestions from different government and private organizations (Council of Scientific and Industrial Research [CSIR], New Delhi, India; Defence Research and Development Organization–Naval Research Board [DRDO–NRB], New Delhi, India; Defence Research and Development Organization–Aeronautical Defence Establishment, [DRDO–ADE], Bengaluru, India; Department of Science and Technology [DST], New Delhi, India; Tata Steels, Jamshedpur, India; Indian Institute of Technologies, Jadavpur University, Kolkata, India; Indian Association for the Cultivation of Science [IACS], Kolkata, India; and Institute of Minerals and Materials Technology [IMMT], Bhubaneswar, India; National Aerospace Laboratories, Bengaluru, India; and the National Metallurgical Laboratory, Jamshedpur, India). We are indebted to entire budding and beautiful Composite Materials Group here at the National Institute of Technology, Rourkela, India.

Authors

Bankim Chandra Ray has been working at the National Institute of Technology, Rourkela, India since 1989. His current designation is Professor and Head of Composite Materials Group and Dean of Faculty Welfare. He has been awarded a PhD from the Indian Institute of Technology, Kharagpur, India in 1993. He had started working on "Environmental Degradation of FRP Composites" from his PhD tenure and has been working continuously on the "Environmental Durability of FRP Composites" for about the last 30 years. He has been associated with several societies, such as the Indian Institute of Metals, Indian National Academy of Engineering, and many other government and private organizations for providing a global platform for FRP composite materials for various structural applications. He is currently serving in the coveted role of an advisor of New Materials Business (FRP Composites) at Tata Steel. He has more than 130 publications in reputed international journals.

Rajesh Kumar Prusty has been working as an Assistant Professor at the National Institute of Technology, Rourkela, India since 2014 after completing his Masters in Engineering from the Indian Institute of Science, Bangalore, India (Gold Medalist). He had secured All India Rank 1 in Graduate Aptitude Test in Engineering (GATE) in 2011 (Metallurgical Engineering). He has been working with Professor Bankim Chandra Ray in FRP Composite Lab at NIT, Rourkela for the past 4 years. He was awarded PhD in 2017, and the topic of his PhD thesis was Implication of CNT on Environmental Durability of FRP Composites based on assessment of Microstructural Features and Mechanical Properties. His focus of research is to Develop advanced FRP Composites for various applications by evaluating their short- and long-term performance. He teaches nanostructured materials and composite materials to the undergraduate and postgraduate students.

Dinesh Kumar Rathore has completed PhD in the field of mechanical behavior of hybrid FRP composites under different elevated temperatures under the supervision of Professor Bankim Chandra Ray in the year 2017. He has authored and co-authored about 20 high impact research and review articles. Currently, he is working as an Assistant Professor at the Kalinga Institute of Industrial Technology (KIIT, Deemed to be University), Bhubaneswar, India. He is teaching mechanics of materials and materials science and engineering to undergraduate students. Dr. Rathore has been awarded best presentation awards at the Indian Institute of Technology, Bombay, India and the National Institute of Technology, Rourkela, India for his PhD work.

1

Introduction

1.1 Applications of Advanced Structural Fiber-Reinforced Polymer Composites

Mechanical properties such as high strength-to-weight ratio, corrosion resistance, good impact resistance, better damping, and fatigue properties enable the use of polymeric composites in wide-ranging fields. Widespread application range of fiber-reinforced polymer (FRP) composites covers nearly every variety of advanced engineering structures. They are used in various parts of airplanes, choppers, spaceships, naval vessels, offshore platforms, as well as in automotive industries, chemical processing equipment, and civil infrastructure such as buildings and bridges [1]. Some of the advanced modern day applications of FRP composite have been provided below to have a feeling of their current position and importance.

1.1.1 Aerospace

When it comes to the aerospace industry, weight is everything. As engineers have constantly tried to enhance the lift-to-weight ratios, here, composites play a major role in weight reduction. Composite materials are the potential and successful alternatives to other materials in aerospace applications because of their appreciable strength and superior physical properties in the different environments. Composites are versatile, such that they can be used for both structural applications and components for a wide range of aerospace vehicles, such as aircrafts, gliders, space shuttles, or hot air balloons. Their application ranges from the complete body of an aircraft to components, such as blades, propellers, seats, and wings.

The applications of FRP composites in the aviation or aerospace industries received serious attention when boron-reinforced epoxy composites were employed as the material for the skins of the components used at the tail of the U.S. F14 and F15 fighters [2]. The framework of an AV-8B, a vertical and short take-off and landing aircraft introduced in 1982, constitutes around 25% by weight of carbon fiber-reinforced polymer (CFRP) composite. In the Boeing line of airplanes, a Boeing 787 (Figure 1.1) is one of the most advanced

FIGURE 1.1
Use of various types of materials in Boeing 787. (From AERO-Boeing 787 from the Ground Up [Online], Available: http://www.boeing.com/commercial/aeromagazine/articles/qtr_4_06/article_04_2.html, 2017.)

planes in the world, revolutionizing air flight in terms of traveler comfort and traveler-friendly solutions. One of the greatest challenges in aviation is efficiently overcoming gravity; hence, they have made half of the Boeing 787 with carbon fiber composites.

In the Airbus series, the A350 XWB's wing (Figure 1.2) comprises of the CFRP composites, including both the covers of the wings. With the dimension of 32 m long and 6 m wide, these are among the largest single aviation components ever made from carbon fiber composites. As it makes the airframe stronger and tougher than those of the conventional airplane frame, along with this, the weight reduction reduces the fuel usage and also increases the carrying capacity of the aircraft.

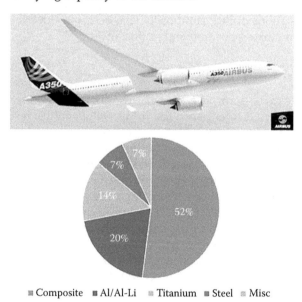

FIGURE 1.2
Use of various types of materials in Airbus A350. (From Airbus A350: Composites on Trial Part I [Online], Available: http://scribol.com/technology/aviation/airbus-a350-composites-on-trial-part-i/, 2017.)

Apart from daily operating costs, the usage of the composite in the aviation industry also simplifies the aircraft maintenance program by the reduction in component counts and corrosion. In aircraft construction, engineers ensure that if there are any possibilities to reduce operating costs, it should be explored and exploited wherever possible. When it comes to aerospace, composite material will continue to exist for a long time and become accepted as normal in the future.

1.1.2 Automotive and Railways

In the automotive industries, the composite materials are used to improve fuel efficiency by reducing the weight of the vehicle without having any significant change in the strength of the component, safety, and crashworthiness, to enhance the styling of the outer body, and to provide the aerodynamic design. While luxury vehicles or automobiles have used composites for different purposes for ages, there is currently a trend toward mid-price models that include many applications of composites. CFRP composites are being used in everything from body parts to brakes [5]. The BMW i8 goes from 0 to 100 km/h in less than 5 s and has consumption of fuel less than 3 L per 100 km [6]. The BMW i8 (Figure 1.3) will feature carbon fiber in the body panels as well as the passenger cell.

To compensate for the additional weight of the electric battery, the passenger compartment has been made up of CFRP composite, which is around 50% lighter than steel and 30% lighter than aluminum, with more or less the same material properties. Carbon fiber is, therefore, a crucial component that makes the BMW i8 different than any other automobile [8].

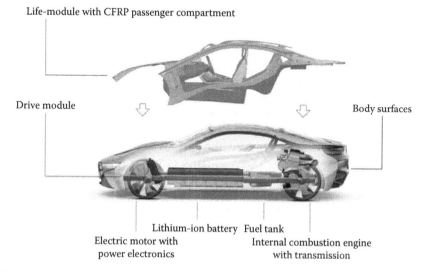

FIGURE 1.3
Extensive use of CFRP composite in BMW i8. (From BMW i design DNA-Car Body Design.)

FIGURE 1.4
FRP composite in heavy load bridge applications. (From Industry Application-Bridge Installations [Online], Available: http://www.tietek.net/bridge_installations.asp, 2017.)

In railway sectors also, due to higher specific properties, FRP composites have gained significant attention these days. After successful testing at its laboratory and also at the Tietek's manufacturing plant in the United States, composite sleepers (Figure 1.4) in 10 railway zones are being laid down across the country by Indian Railways [9].

1.1.3 Marine

The marine industries are very well known in the composites industry. Manufacturers use composites in about every components or in body parts from naval vessels to ferries and boats to submarines. However, even though glass FRP is the widely used material for marine applications, CFRP is gaining attention in some critical applications currently [11].

Gold Coast Yachts has achieved the improved weight and engineering specifications for the B53 high-performance racer/cruiser. The new B53 catamaran (Figure 1.5), a 53-ft., high-performance yacht, is made entirely of CFRP composite [12].

Marine application covers military surveillance ships. The government of Mozambique needed an offshore patrol vessel that has better fuel efficiency with fast acceleration for quick pursuits, which is fulfilled by a 143-ft. long trimaran with hulls made of glass fiber and epoxy resins. The vessel's high-load areas, such as the arms that connect the three hulls, are made with the CFRP composite.

FIGURE 1.5
Use of carbon fiber for B53 catamaran. (From Bieker 53-Exciting high performance cruiser-Cruisers & Sailing Forums [Online], Available: http://www.cruisersforum.com/forums/f48/bieker-53-exciting-high-performance-cruiser-144229.html, 2017.)

The new composite molds applied by the Boston Whaler launched the Outrage 420 in 2014. The Outrage 420 was named as the Boat of the Year, 2015 by a boating magazine [14]. The blades, as long as 300 ft., are formed in a 2-part closed mold made from strong, but lightweight composites.

1.1.4 Infrastructure

FRP composite is a potential alternative for reinforced concrete and steel in bridges, buildings, and for other structural applications. Composites are preferably used for corrosion-resistant pipes or tubes. Currently, many bridge decks have been constructed from FRP composites.

Pont y Ddraig (Dragon Bridge), Rhyl, United Kingdom (Figure 1.6)—a lifting bridge, has been constructed from lightweight composites to avoid wastage of energy. The design is comprised of a central mast-like structure and caisson with two wing-like FRP spans that rise at the same time for letting the boat traffic pass. Each of the two bridge decks measures 32 m long and just 6 m at their widest point, creating a magnificent outlook, particularly when the wings are raised [15].

Wolf Trap Pedestrian Bridge in Vienna, Austria (Figure 1.7) is a lightweight bridge with a deck that provides simple installation with very little traffic disruption. The components employed in this installation were 55 prefabricated FRP bridge deck panels, connected to a steel truss using mechanical fasteners and Z-clips. An epoxy aggregate non-slip coating is applied on the surface of the bridge deck. As FRP was used due to its high strength-to-weight

FIGURE 1.6
Rhyl Pont y Ddraig lifting bridge. (From Composite materials: enter the dragon, *The Manufacturer* [Online], Available: https://www.themanufacturer.com/articles/composite-materials-enter-the-dragon/, 2017.)

FIGURE 1.7
Wolf Trap Pedestrian Bridge in Vienna. (From FRP Transportation Infrastructure [Online], Available: http://www.compositesinfrastructure.org/case-studies/wolf-trap-pedestrian-bridge-vienna-va/, 2017.)

ratio, apart from the pedestrians, the bridge has the strength to support an emergency vehicle if necessary and FRP also makes it easier to lift and move because the bridge deck is around 80% lighter than a conventional deck made up of gravel and cement [16].

1.1.5 Some Other Special Applications

1.1.5.1 NASA and Boeing Build and Test All-Composite Cryogenic Tank

NASA and Boeing have now successfully built and tested an all-composite, CFRP 5.5-m diameter tank (Figure 1.8). Being one of the biggest and lightest cryogenic liquid hydrogen fuel tanks manufactured in history, the composite cryotank is also a step forward to the planned 8.4-m tank that aims at the reduction of the weight of rocket tanks by 30% and cuts launch expenditures by a minimum of 25% [17].

1.1.5.2 Liquefied Petroleum Gas Cylinders

Composite cylinders (Figure 1.9) have gained huge interest and appreciation across the globe due to their various features, such as being extremely light, corrosion proof, ultraviolet (UV) resistant, and most importantly are 100% explosion proof.

1.1.5.3 Wind Turbine

Composite wind turbines provide electricity that helps power monuments. Wind turbines (Figure 1.10) put on the scaffolding 400 ft. from the ground

FIGURE 1.8
Cryogenic liquid hydrogen fuel storage tank. (From All-Composite Cryogenic Tank Build and Tested by NASA and Boeing, *CompositesLab.*)

FIGURE 1.9
Light weight FRP cylinder for LPG. (From Flaunt your LPG cylinder [Online], Available: http://www.merinews.com/article/flaunt-your-lpg-cylinder/15858189.shtml, 2017.)

FIGURE 1.10
GFRP wind turbine installed in Eiffel tower. (From The Eiffel Tower Now Has Two Eco-Friendly GFRP Turbines, *CompositesLab* [Online], Available: http://compositeslab.com/eiffel-tower-now-has-two-eco-friendly-gfrp-turbines/, 2017.)

level have made the Eiffel Tower a little more eco-friendly. The spinning of the turbines glass fiber composite blades produces enough wind energy to handle the electrical needs of the nearby souvenir shop and restaurants. The turbines supply 10,000 kWh of electricity per year, enough to power a typical American family's home [20].

1.2 Constituents of Fiber-Reinforced Polymer Composites

1.2.1 Polymer Matrix

The primary aspect of determining an appropriate matrix is with consideration to its basic mechanical properties, which includes its tensile strength, tensile modulus, and fracture toughness. Additional properties required to determine polymer matrices are their dimensional stability at increased temperatures (determined by T_g) and their resistance to solvents and moisture. Thermoset polymers have an advantage over thermoplastic polymers with respect to their thermal stability and chemical resistance, whereas thermoplastic polymers have a decided advantage over the former in terms of high impact strength and fracture toughness. General information on some of the thermosetting and thermoplastic polymers has been listed below.

1.2.1.1 Thermosetting Matrix

Epoxy: The structure of an epoxide group is shown in (Figure 1.11).

The curing of an amino–epoxy system involves three main reactions as shown in Figure 1.12. The density of cast epoxy resin at room temperature is 1.2–1.3 g/cm³, with a recorded tensile strength of 55–130 MPa, tensile modulus of 2.75 GPa, and Poisson's ratio in the range of 0.2–0.33 [21].

It is the cross-link density, which majorly determines the properties of the cured epoxy resin. The chemical structure of the initial

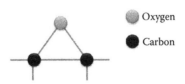

FIGURE 1.11
Structure of an epoxide group.

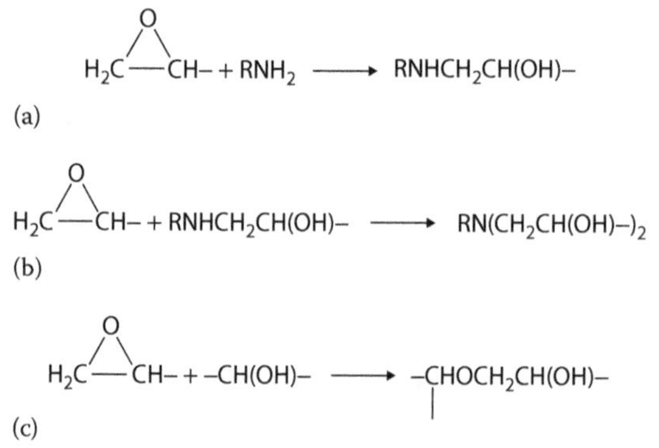

(a)

(b)

(c)

FIGURE 1.12
Schematic depiction of epoxy curing, (a) primary amine addition, (b) secondary amine addition, and (c) etherification. (From Vyazovkin, S. and Sbirrazzuoli, N., *Macromolecules*, 29, 1867–1873, 1996.)

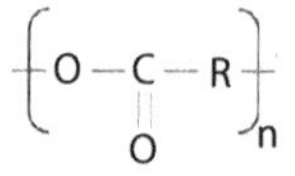

FIGURE 1.13
A simple polyester structure.

epoxy group, the significance of the curing agent, and reaction conditions are some of the factors that determine the cross-link density.

Polyester: The starting material here is an unsaturated double-bonded polyester resin. The curing reaction is conducted by adding quantities of a chosen catalyst, such as organic peroxide or an aliphatic azo compound to the designated liquid mix. The basic building block of polyester has been shown in Figure 1.13.

The density of cast thermoset polymer resins at room temperature is 1.1–1.43 g/cm^3, with a tensile strength of 34.5–103.5 MPa, and tensile modulus of 2.1–3.45 GPa [21].

Vinyl ester: Here, the starting material is a vinyl ester resin created by the reaction of an unsaturated carboxylic acid, such as methacrylic or acrylic acid, and an epoxy. The double bonds are present only at the ends of the vinyl ester molecule. The presence of –OH groups along the length leads to excellent wet-out and adhesive properties (Figure 1.14).

The density of cast vinyl ester resins at room temperature is 1.12–1.32 g/cm^3, with a tensile strength in the range of 73–81 MPa, and tensile modulus in the range of 3–3.5 GPa [21].

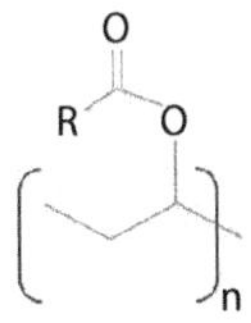

FIGURE 1.14
A typical vinyl ester structure.

1.2.1.2 Thermoplastic Matrix

Identity property of these polymers is the presence of rigid aromatic rings, giving them higher glass transition temperature and a brilliant dimensional stability.

  Polyether ether ketone: Polyether ether ketone is a semi-crystalline polymer with a linear aromatic chain following the repeating unit in its molecules (Figure 1.15). The melt processing of the polyphenylene sulfide requires heating the said polymer in the temperature range of 300°C–345°C. The continuous use temperature is 240°C. It has an excellent chemical resistance.

  Polyether ether ketone shows high fracture toughness, much higher than that of epoxies. Also, it shows a low water absorption capacity and, being semi-crystalline, does not dissolve in common salts.

  Polyphenylene sulfide: Polyphenylene sulfide is a semi-crystalline polymer, showing a glass transition temperature of 85°C, with the following repeating unit in the molecules (Figure 1.16). It has a high tensile strain-to-failure performance (50%–100%) and a brilliant hydrolytic stability under hot–wet conditions (e.g., in steam). Although polysulfone has a good resistance to mineral acids, alkalis,

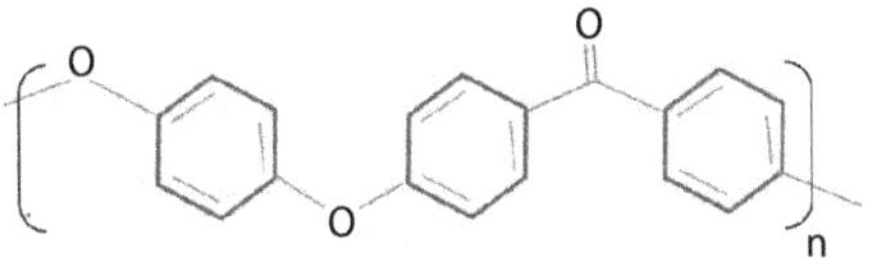

FIGURE 1.15
Structure of polyether ether ketone (PEEK).

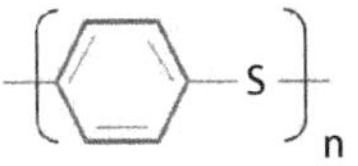

FIGURE 1.16
Structure of polyphenylene sulfide.

and salt solutions, it will, however, swell, stress-crack, or dissolve in polar organic solvents, such as ketones, chlorinated hydrocarbons, and aromatic hydrocarbons.

Its comparatively low glass transition temperature is because of the flexible sulfide linkage between the aromatic rings.

1.2.2 Reinforcements

The manufacturing of a composite involves the incorporation of a large number of fibers in a thin layer of matrix to form a lamina (ply). Fibers can be arranged in either unidirectional (i.e., one direction) or in bidirectional (i.e., in two directions, usually perpendicular to each other) orientations (Figure 1.17).
A lamina can also be built using discontinuous (short) fibers in a matrix. Reinforcements can also be present in the form of flakes, whiskers, particulates, short fibers, continuous fibers, or sheets. The extensive use of fibers as reinforcement materials arises from the fact that they have a small diameter, high aspect ratio, and a very high degree of flexibility. Common types of fibers are glass, carbon, aramid and Kevlar fibers, and natural fibers, which include jute, flax, and hemp and banana fiber. Ceramic fibers, like SiC and Al_2O_3 find their use in high-temperature applications.

1.2.2.1 Glass Fibers

Glass fibers are classed into several types: E glass fibers, where E stands for high electrical insulation; S glass fibers, where S stands for high silica content; and C glass fibers, where C denotes corrosion resistance. Glass fibers are observed to possess low density (2.55 g/cm^3), but high strength 1750 MPa with a moderate Young's modulus of 70 GPa. Thus, while the strength-to-weight ratio is extremely high, the modulus-to-weight ratio is only moderate. Glass fibers experience a decreased strength due to moisture absorption, along with a condition known as static fatigue, whereby subjecting the fibers to a continuous load for an extended period of time can lead to substantial crack growth.

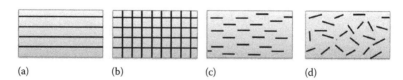

(a)   (b)   (c)   (d)

FIGURE 1.17
Principal building blocks in fiber-reinforced composites, (a) unidirectional continuous, (b) bidirectional continuous, (c) unidirectional discontinuous, and (d) random discontinuous.

1.2.2.2 Carbon Fibers

Carbon fibers have an extremely low density of 2.268 g/cm³. In general, the low-modulus carbon fibers have a relatively lower density, lower cost, higher tensile and compressive strengths, and higher tensile strains-to-failure than the high-modulus carbon fibers. The advantages of carbon fibers are their high strength-to-weight ratios, as well as tensile modulus-to-weight ratios, extremely low coefficient of the linear thermal expansion, high fatigue strength, and high thermal stability. The evident disadvantages are their low strain-to-failure, low impact resistance, and their high electrical conductivity, which may cause "shorting" in unprotected electrical machinery.

1.2.2.3 Aramid Fibers

Aramid fibers are excellently crystalline aromatic polyamide fibers (Figure 1.18) that have the lowest density and the highest of the tensile strength-to-weight ratio among the current reinforcing fibers. Kevlar 49 is the most well known trade name of one of the aramid fibers available widely in the market. The molecular structure of aramid fibers, such as Kevlar 49 fibers, is illustrated in Figure 1.18. The observed aromatic ring gives it a higher chain stiffness (modulus), as well as a better chemical and thermal stability over other commercial organic fibers, such as those of nylons.

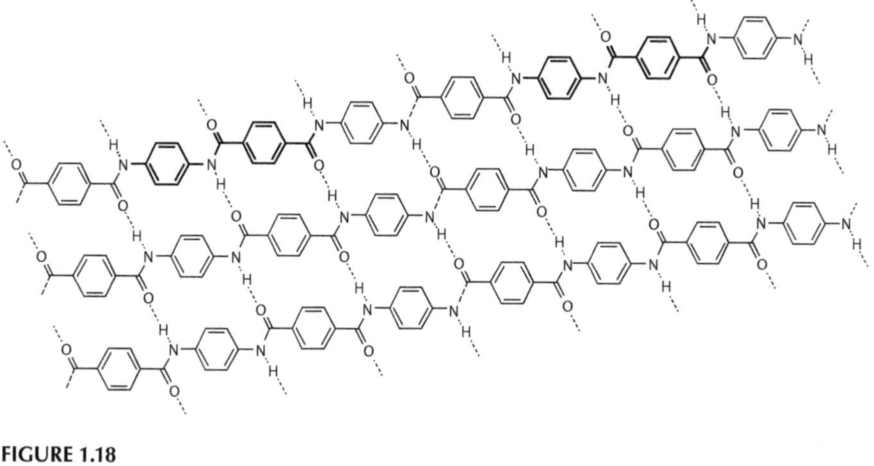

FIGURE 1.18
Molecular structure of the Kevlar 49 fiber. (From chempolymerproject-Kevlar-E-nydw [Online], Available: https://chempolymerproject.wikispaces.com/Kevlar-E-nydw, 2017.)

1.2.2.4 Boron Fiber

The most evident feature of boron fibers is their extremely high tensile modulus, which is in the range of 379–414 GPa. In addition to their relatively large diameter, boron fibers offer an excellent resistance to buckling, which in turn attributes to high compressive strength for boron fiber-reinforced composites. The significant disadvantage of boron fibers is their high cost, which is even higher than that of many forms of other carbon fibers. For this reason, its application is at present restricted to a few aerospace applications.

Theoretical calculations for its strength, modulus, and other properties depend on the fiber volume in the matrix and can be calculated from the following equations:

$$vf = \frac{\dfrac{w_f}{\rho_f}}{\left(\dfrac{w_f}{\rho_f}\right) + \left(\dfrac{w_m}{\rho_m}\right)} \tag{1.1}$$

$$\rho_c = \frac{1}{\left(\dfrac{w_f}{\rho_f}\right) + \left(\dfrac{w_m}{\rho_m}\right)}, \tag{1.2}$$

where:
  w_f is the fiber weight fraction (same as fiber mass fraction)
  w_m is the matrix weight fraction (same as matrix mass fraction) and is
     equal to $(1 - w_f)$
  ρ_m is the density of the matrix
  ρ_f is the density of the fiber

In terms of volume fraction, the composite density can be written as follows:

$$\rho_c = \rho_f v_f + \rho_m v_m, \tag{1.3}$$

where:
  v_f is the fiber volume fraction
  v_m is the matrix volume fraction. Here, $v_m = 1 - v_f$

The purpose of the reinforcement of the matrix is to produce high strength and stiffness of the composite.

1.2.3 Interface

Interface is the prominent region, which determines, to a large extent, the set of properties of all heterogeneous systems, including that of composite materials. A typical composite consists of a continuous phase (the matrix)

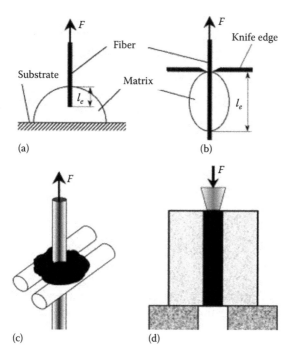

FIGURE 1.19
External load applied to the fiber, (a) fiber pull-out test, (b) micro bond test, (c) three fiber test, and (d) push out test. (From Zhandarov, S. and Mäder, E., *Compos. Sci. Technol.*, 65, 149–160, 2005.)

and inclusions of the other phase(s) distributed within it (in the form of particles, fibers, or elements having other shape) [24]. From the chemical point of view, the strength of the interfacial interaction depends on the surface concentration of the interfacial bonds and the bond energies.

Quite a number of micromechanical tests have been developed to determine the interfacial strength between a fiber and a matrix. In one group, an external load is directly put to the fiber. These are shown in Figure 1.19.

In the second case, the matrix is then put through an externally loaded-fragmentation test (Figure 1.20).

Normally, the quality of an interfacial bond is characterized by the calculation of the apparent interfacial shear strength according to the definition:

$$\tau_{app} = \frac{F_{max}}{\pi d_f l_e},$$ (1.4)

where:
d_f is the fiber diameter
l_e is the embedded length

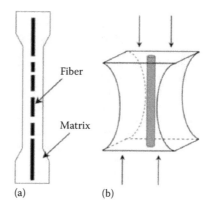

FIGURE 1.20
External load applied to the matrix, (a) fragmentation test, (b) Broutman test. (From Zhandarov, S. and Mäder, E., *Compos. Sci. Technol.*, 65, 149–160, 2005.)

The quantitative approach of fiber-matrix interface provides a more sufficient result, and, therefore, multiple models of the stress-distribution and the interfacial failure in fiber-matrix composites have been proposed [24].

1.2.4 Sizing

The sizing is a mixture of various chemicals, usually (but not necessarily) diluted in water, which fiber and fabric producers coat ("size") their fibers with (Figure 1.21). Sizing [25] follows a very complex band of ingredients: one or several polymeric components, a coupling agent, lubricant, and a range of additives (surfactants, plasticizers, anti-static agents, adhesion promoters, anti-foams, rheology modifiers, etc.).

Sizing plays a role in production, processing of the fiber, and interfacial bonding between the fiber and the matrix. The function of a sizing is to form either a temporary or often a permanent interface between the fiber and the matrix, and, therefore, every matrix requires a different sizing chemistry.

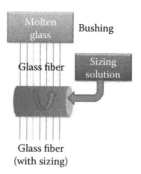

FIGURE 1.21
Schematic representation of sizing applications during fiberglass manufacturing.

Glass fiber sizing is not a single chemical compound, but a mixture of several complex chemistries, each of which contributes to the sizing's overall performance [26]. The two major constituents are the film former and the coupling agent. The function of a film former is to protect and lubricate the fiber and hold fibers together prior to molding, along with promoting their separation when in contact with resin, ensuring wet out of all the filaments. The coupling agent, almost always an alkoxysilane compound, serves primarily to bond the fiber to the matrix resin. The major use of silanes in this case arises from their ability to bond two highly dissimilar metals. Silanes have a silicon end that bonds well to glass and an opposing organic end that bonds well to resins.

1.3 Fabrication Techniques of Fiber-Reinforced Polymer Composites

Several techniques for manufacturing of FRP composites have been listed below.

1.3.1 Hand Lay-Up Method

The hand lay-up method is the simplest method and is also termed as wet lay-up method as shown in Figure 1.21. Here, the reinforcement is manually placed in the mold and subsequently resin is applied. The mold can be as simple as a flat sheet. Release agents (wax, silicone, and release paper) are applied for easy removal of the part. A brush and a roller are used to impregnate the fibers with the resin (Figure 1.22).

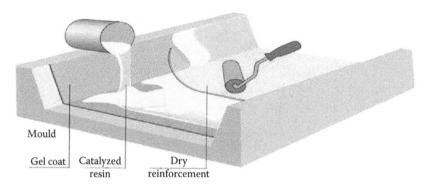

Mould

Gel coat | Catalyzed resin | Dry reinforcement

FIGURE 1.22
A simple hand lay-up method of composite fabrication. (From Nuplex-Hand Lay-Up [Online], Available: http://www.nuplex.com/composites/processes/hand-lay-up, 2017.)

Initially the mold is treated with a releasing spray (for easy removal later on) followed by the application of a thin layer of polymer. Then a fiber in the form of a mat or cloth is placed on it and gentle rolling is done to impregnate the fiber mat with the earlier applied polymer resin by a roller. This process is then repeated as per the required thickness of the composite. The curing is then done to harden the polymer, and finally the composite is removed from the mold. Applications of hand lay-up method are in boat hulls, swimming pools, and large container trunks.

1.3.2 Spray-Up Method

Here in the spray-up method, fiber is chopped in a handheld gun and fed into a spray of catalyzed resin directed at the mold. The deposited materials are cured in standard atmospheric conditions. Primarily polyester resins with glass roving fibers are generally applied by this technique (Figure 1.23).

1.3.3 Pultrusion

Pultrusion is a continuous manufacturing process. This process is conducive for manufacturing of constant cross-section shapes of any length. Here, filaments are pulled from a creel passed through a resin bath and extruded through a heated die. The die then completes the impregnation of the fiber. It further controls the resin content and cures the material to its final shape. The cured profile is then automatically brought to the desired constant length (Figure 1.24). The main advantage of pultrusion is that it is a very fast, as well as an economic way of impregnating and curing materials. Also, the laminates show good structural properties since the fibers have a straight profile.

Composites fabricated through this process find applications in beams, girders, and other such construction purposes.

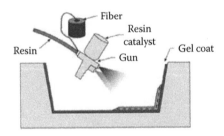

FIGURE 1.23
Schematic diagram for spray-up method. (From Advanced Composites Manufacturing, Carbon Fiber Parts Engineering Processes|wet layup, Vacuum resin infusion, Vacuum Resin transfer molding, injection molding [Online], Available: http://www.carbonfiberglass.com/ Composites-Manufacturing/Composites-Manufacturing-Processes.html, 2017.)

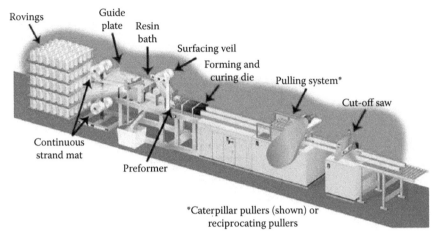

FIGURE 1.24
Schematic diagram of pultrusion method of composite fabrication. (From The Pultrusion Process, Strongwell [Online], Available: http://www.strongwell.com/about/the-pultrusion-process/, 2017.)

1.3.4 Filament Winding

The filament winding process is applied for hollow components having general circular and oval shaped components, such as pipes and tanks. Here, the fibers tow pass through a resin bath and are then wound into a mandrel in a variety of orientations, that are controlled by a fiber-feeding mechanism, along with the speed of rotation of the mandrel (Figure 1.25).

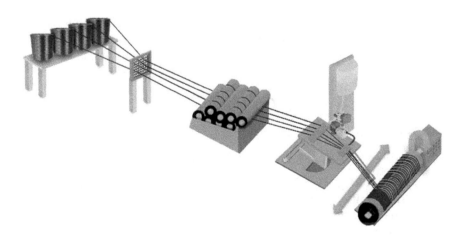

FIGURE 1.25
Schematic diagram of Filament winding method. (From Filament Winding Process Application-Basalt Fiber Tech [Online], Available: http://www.basaltft.com/app/fw.htm, 2017.)

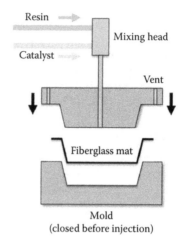

FIGURE 1.26
Resin-transfer molding. (From Resin Transfer Molding, Orenco composites [Online], Available: http://orencocomposites.com/processes/rtm/, 2017.)

1.3.5 Resin Transfer Molding

Resin transfer molding is a closed molding process. Here, resin is transferred over the placed reinforcement. The reinforcement is in the shape of a strand mat or a woven mat placed neatly on the lower half of the mold. Release gel is applied to easily remove the composite from the mold, and the mold is properly closed and clamped. The resin is pumped into the mold through ports. Air is displaced through vents located in the mold. Once curing is completed, the mold is opened and the composite is removed. Matrices and fibers of almost all kinds can be applied in this form of fabrication process. Mineral fillers may also be applied to improve surface finish (Figure 1.26). The major disadvantage of this process is that mold cavity limits the size of the composite.

1.4 Different In-Service Environments

Exposing the FRP composite materials for long durations to different environmental conditions will finally have an irreversible effect in initial properties (mechanical, physical, or chemical) of the material and hence limiting the operational life.

1.4.1 Environmental Factors

Change in properties over time in the material is called ageing. Ageing depends highly on material characteristics and environment. An ageing environment can be static or dynamic in nature. The study of ageing due to different environmental conditions is very essential for today's application in different sectors. Some parameters, which affect the ageing behavior of the materials due to the in-service environment, are discussed below.

1.4.1.1 Temperature

As majorly as the ageing process depends on temperature, it is the most important parameter to be considered while studying the ageing process of material. At different temperatures, most of the polymers have distinct material phases. Physical ageing will take place when a polymer is rapidly cooled below its second order transition temperature below the melting temperature (T_g), which is also known as glass transition temperature. Hence, to maintain thermodynamic equilibrium, there is a resultant change in different parameters like enthalpy, free volume energy, and entropy of the polymer and will produce appreciable changes in the mechanical properties.

1.4.1.2 Water or Moisture Absorption (Humidity)

Moisture absorption can affect the thermoplastic resins to a large extent. The water absorption can degrade the materials in different ways. First, it will lead to an increase in chain mobility (plasticizing effect), which is visible at the macroscale resulting in the degradation of elastic modulus, the increment in the elongation at break, and the lowering of glass transition temperature [32].

1.4.1.3 Ultraviolet Radiation and Other High Energy Radiations

UV radiation affects the properties of composites by affecting the polymeric resin used in the composite, as the photo degradation being a surface mechanism is confined to loss of mechanical properties of resin (fiber is not degraded), this signifies that the property of resin is crucial. Chemical changes caused by UV exposure are due to a complex set of processes, which involves both UV and oxygen effects on the material. The basic process for photo initiated degradation is similar for all of the polymeric materials. Photons harm the linking of bonds by reacting with the molecular chains that constitute the polymeric structures. The wavelengths which are shorter have higher photon energies, are more easily absorbed in polymeric materials, and have a higher potential to disrupt chemical bonds of the material [33].

1.4.1.4 Low Earth Orbit

Molecular oxygen is stable, but atomic oxygen is very reactive in nature and can affect the materials in contact with it. This type of environmental factor affects many aircraft, helicopter components, or outer body parts.

1.4.1.5 Acid Rain

Attacks due to chemicals are very critical as they harm the materials and are a critical parameter to be taken into consideration. Chemical ageing is similar to the irreversible changes that take place in the polymeric chains by mechanisms such as cross-linking or chain breakage.

1.4.2 *In-Situ* Environments during Various Applications of Fiber-Reinforced Polymer Composites

1.4.2.1 Ageing of Composites in Marine Applications

Even before the advancement of composite materials, these are used for marine applications. Using composite materials as the building material for the marine vessels will not only provide corrosion resistance, but also, less water absorption, lighter weight body of the vessel, and stylish outer finish. Composition of the seawater and ambient temperatures are two crucial factors that will influence the ageing behavior of the composites used in marine application. Mechanisms involved in wet ageing of the polyester composites are as follows: physical degradation due to the interfacial debonding and delamination, plasticization and swelling, and chemical degradation, which involves the hydrolysis of the matrix and fiber degradation under certain conditions. The temperature usually varies from 5°C to 50°C for marine applications. Composites used for marine applications generally have lower volume fractions of fiber (30%–60%), are fabricated by hand lay-up, and impregnation using low temperature cure resins [34].

1.4.2.2 Ageing of Composites in Underwater Applications

Underwater applications are generally concerned with military applications and are used for deep-sea inspection and study of marine life. From the last decade, demand for composite materials in underwater applications has been rising rapidly. Many factors should be kept in mind while designing an underwater vessel, as for example: material of light weight is of importance in submarine structures, should withstand high pressure, and also should have appreciable corrosion resistance. The FRPs are selected due to their resistance to seawater and good specific mechanical properties, such as damping of waves and nonmagnetic behavior.

There are many cases where there are both the change in mechanical properties and water absorption kinetics of materials, they are studied like:

1. The properties of composite tubes made up of glass/epoxy used for cooling water applications are evaluated. At different temperatures and internal pressures, the long term ageing of tubes and different resin hardeners and the influence of liners were studied. After analyzing the results, we conclude that the mechanical properties of material such as flexural strength and interlaminar shear strength are significantly affected by the hydrolysis.

2. Syntactic foam is used in underwater applications. In specific applications, there is a need of materials with properties, as these properties can be attained by the application of composites. Hence, it offers a wide range of use for composites. Syntactic foams are used as a core of composite sandwich structures and are attractive due to the properties like thermal insulation and buoyancy [35].

1.4.2.3 Ageing of Composites in Aerospace Applications

Materials used in aerospace applications have to withstand different harsh environments. In aerospace application, the selection of material depends on different factors like: humidity, temperature, atmospheric pressure, and different radiations to which the material is subjected. Composites are more resistant when compared with metal to fatigue because of repeated take off/landing processes, which cuts down the expensive inspections over the aircraft's lifespan. When compared with metals, it does not corrode. Hence, both fatigue and corrosion resistance combined are a major problem that is faced by airplanes and can be overcome easily by the use of composite in the body components of the aircraft.

Low counts of the components and reduction in the weight of the components are also two of the major advantages of application of composites in aerospace. Thermal and electrical shock resistance also is crucial considering the environmental conditions. Heat and impact resistant thermoplastic composites are used for fixed wing leading edges, engine inlets, and throughout a variety of structural applications ranging from clips and brackets to engine pylon doors and control surfaces in the aerospace industry.

1.4.2.4 Ageing of Composites in Oil and Chemical Industries

With the advancement in the oil and chemical sector, there has been a rise in the requirement for the materials performing in different conditions like high temperature and pressure, corrosive environments, and improved strength with light weight. Composites are used for containing and transportation of the chemicals. FRP composites are used heavily in the industries

FIGURE 1.27
TenCate compressed gas cylinders. (From Oil and gas-TenCate Advanced Composites, Advanced composites [Online], Available: https://www.tencatecomposites.com/markets/industrial/oil-and-gas, 2017.)

that manufacture concentrated acids, chlorine, and chlorate. These materials are also being used as components for desulfurization plants like scrubbers and ducts.

In the oil and gas sector, we require a material, which should be light weight and corrosion resistant, and we require a high level of weight reduction and reliability, which is fulfilled by composites. TenCate's thermosets overwrap the compressed gas cylinders (Figure 1.27) used for a variety of command and control actions on valves. Composites in this application provide high levels of weight reduction and reliability [36].

1.4.2.5 Ageing of Composites in Structural Applications

Over many decades, FRPs have been used in a wide range of applications in construction and structural sectors like bridges, gratings, pipes, walkways, tanks, and slabs. In these applications, the material is exposed to one or more environmental influences. The composite materials are used because of their high durability, less deterioration caused due to weathering, lightness, as well as structural strengthening properties and are used for stand-alone components.

The evolution of the foldable FRP Cargoshell vessel and the outer skin of the Uyllander Bridge (Figure 1.28) near Diemen, the Netherlands is a perfect example of the structural application [37]. So for structural applications, we required materials that have high durability and are resistant to electrical and thermal shock due to climatic changes in the environment.

FIGURE 1.28
Uyllander Bridge, Diemen (the Netherlands). (From Uyllander bridge [Online], Available: http://europe.arcelormittal.com/fol_uyllander, 2017.)

References

1. L. C. Hollaway, A review of the present and future utilisation of FRP composites in the civil infrastructure with reference to their important in-service properties, *Constr Build Mater*, 24(12): 2419–2445, 2010.
2. A. Quilter, Composites in aerospace applications, *IHS White Paper*, 444(1), 2001.
3. AERO-Boeing 787 from the Ground Up [Online]. Available: http://www.boeing.com/commercial/aeromagazine/articles/qtr_4_06/article_04_2.html. [Accessed: October 11, 2017].
4. Airbus A350: Composites on Trial Part I [Online]. Available: http://scribol.com/technology/aviation/airbus-a350-composites-on-trial-part-i/. [Accessed: October 11, 2017].
5. Composites in Automotive Industry-Applications, *CompositesLab*. http://compositeslab.com/where-are-composites-used/automotive-applications/.
6. BMW rolls out prototypes of composites-intensive i3, i8 [Online]. Available: http://www.compositesworld.com/news/bmw-rolls-out-prototypes-of-composites-intensive-i3-i8. [Accessed: September 19, 2017].
7. BMW i design DNA-Car Body Design. http://www.carbodydesign.com/2012/06/bmw-i-design-dna/.
8. BMW i8: Efficiency & dynamics [Online]. Available: https://legacy.bmw.com/com/en/newvehicles/i/i8/2014/showroom/efficiency_dynamics.html. [Accessed: September 19, 2017].

9. Composite Sleepers|Products|Patil Group [Online]. Available: http://www. patilgroup.com/composite-sleepers.html. [Accessed: September 19, 2017].

10. Industry Application-Bridge Installations [Online]. Available: http://www. tietek.net/bridge_installations.asp. [Accessed: October 11, 2017].

11. Composites in Marine Industry, *CompositesLab.* http://compositeslab.com/ where-are-composites-used/marine-applications/.

12. Gold Coast Yachts Uses Custom Carbon Fiber for Sleek Catamaran, *CompositesLab.* http://compositeslab.com/gold-coast-yachts-uses-custom-carbon-fiber-for-sleek-catamaran/.

13. Bieker 53-Exciting high performance cruiser-Cruisers & Sailing Forums [Online]. Available: http://www.cruisersforum.com/forums/f48/bieker-53-exciting-high-performance-cruiser-144229.html. [Accessed: October 11, 2017].

14. Gallery: 2015 Boat of the Year: Boston Whaler 420 Outrage, Boating Magazine [Online]. Available: https://www.boatingmag.com/2015-boat-year-boston-whaler-420-outrage. [Accessed: March 01, 2018].

15. Composite materials: Enter the dragon, *The Manufacturer* [Online]. Available: https://www.themanufacturer.com/articles/composite-materials-enter-the-dragon/. [Accessed: October 11, 2017].

16. FRP Transportation Infrastructure [Online]. Available: http://www. compositesinfrastructure.org/case-studies/wolf-trap-pedestrian-bridge-vienna-va/. [Accessed: September 19, 2017].

17. L. Mohon, Game changing composite cryogenic propellant tank completes testing, *NASA*, May 20, 2015 [Online]. Available: http://www.nasa.gov/content/ nasa-tests-game-changing-composite-cryogenic-fuel-tank_marshall_news. [Accessed: October 11, 2017].

18. All-Composite Cryogenic Tank Build and Tested by NASA and Boeing, *CompositesLab.*

19. Flaunt your LPG cylinder [Online]. Available: http://www.merinews.com/ article/flaunt-your-lpg-cylinder/15858189.shtml. [Accessed: October 2017].

20. The Eiffel tower now has two eco-friendly GFRP turbines, *CompositesLab* [Online]. Available: http://compositeslab.com/eiffel-tower-now-has-two-eco-friendly-gfrp-turbines/. [Accessed: October 11, 2017].

21. P. K. Mallick, *Fiber-Reinforced Composites: Materials, Manufacturing, and Design.* Boca Raton, FL: CRC Press, 2007.

22. S. Vyazovkin and N. Sbirrazzuoli, Mechanism and kinetics of epoxy–amine cure studied by differential scanning calorimetry, *Macromolecules*, 29(6): 1867–1873, 1996.

23. chempolymerproject—Kevlar-E-nydw [Online]. Available: https://chempoly-merproject.wikispaces.com/Kevlar-E-nydw. [Accessed: October 11, 2017].

24. S. Zhandarov and E. Mäder, Characterization of fiber/matrix interface strength: Applicability of different tests, approaches and parameters, *Composites Science and Technology*, 65(1): 149–160, 2005.

25. Composite materials guide: Reinforcements-Sizing Chemistry | NetComposites [Online]. Available: https://netcomposites.com/guide-tools/guide/reinforce-ments/sizing-chemistry. [Accessed: October 11, 2017].

26. K. Mason, Sizing up fiber sizings [Online]. Available: https://www.composites-world.com/articles/sizing-up-fiber-sizings. [Accessed: October 11, 2017].

27. Nuplex-Hand Lay-Up [Online]. Available: http://www.nuplex.com/composites/ processes/hand-lay-up. [Accessed: October 11, 2017].

28. Advanced Composites Manufacturing, Carbon Fiber Parts Engineering Processes|wet layup, Vacuum resin infusion, Vacuum Resin transfer molding, injection molding [Online]. Available: http://www.carbonfiberglass.com/Composites-Manufacturing/Composites-Manufacturing-Processes.html. [Accessed: October 11, 2017].

29. The Pultrusion Process, Strongwell, May 13, 2013. [Online]. Available: http://www.strongwell.com/about/the-pultrusion-process/. [Accessed: October 11, 2017].

30. Filament Winding Process Application-Basalt Fiber Tech [Online]. Available: http://www.basaltft.com/app/fw.htm. [Accessed: October 13, 2017].

31. Resin Transfer Molding, Orenco composites [Online]. Available: http://orencocomposites.com/processes/rtm/. [Accessed: October 11, 2017].

32. M. Broudin et al., Water diffusivity in PA66: Experimental characterization and modeling based on free volume theory, *Eur Polym J*, 67: 326–334, 2015.

33. M. M. Shokrieh and A. Bayat, Effects of ultraviolet radiation on mechanical properties of glass/polyester composites, *J Compos Mater*, 41(20): 2443–2455, 2007.

34. P. Davies, F. Mazeas, and P. Casari, Sea water aging of glass reinforced composites: Shear behaviour and damage modelling, *J Compos Mater*, 35(15): 1343–1372, 2001.

35. Syntactic Foam Thermal Insulation for Ultra-Deepwater Oil and Gas Pipelines [Online]. Available: https://www.researchgate.net/publication/266670238_Syntactic_Foam_Thermal_Insulation_for_Ultra-Deepwater_Oil_and_Gas_Pipelines. [Accessed: September 21, 2017].

36. Oil and gas-TenCate Advanced Composites, Advanced composites [Online]. Available: https://www.tencatecomposites.com/markets/industrial/oil-and-gas. [Accessed: September 22, 2017].

37. Uyllander bridge [Online]. Available: http://europe.arcelormittal.com/fol_uyllander. [Accessed: October 11, 2017].

2

Micro- and Macrocharacterization Techniques

2.1 Introduction

The fundamental mechanical properties of fiber-reinforced composites are assessed by their response under tensile, compressive, and shear loading conditions. These basic material properties are prerequisite for the design and development of any fiber-reinforced polymer (FRP) component. In this chapter, various mechanical, as well as microcharacterization techniques, such as scanning electron microscopy (SEM), atomic force microscopy (AFM), and Fourier transform infrared (FTIR) spectroscopy are presented.

2.2 Static Mechanical Characterization

2.2.1 Tensile Test

The tensile test is one of the fundamental tests used to evaluate the tensile properties, such as tensile modulus, tensile strength, and Poisson's ratio. A flat rectangular specimen can be utilized as per American Society for Testing and Materials (ASTM) D3039 [1]. The end tabs can be made stronger by adhesive bonding of composite strips to ensure the failure within the gauge area. The specimen is fixed in the grips of the fixture and a constant crosshead velocity (2 mm/min) is employed to conduct the test. The longitudinal and transverse strains are measured with the strain gauges. The longitudinal and transverse properties are evaluated by testing 0° and 90° directional laminate, respectively.

2.2.2 Short Beam Shear Test

The short beam shear test is a 3-point flexural test on a specimen with a small span, which promotes failure by interlaminar shear [2]. The shear stress

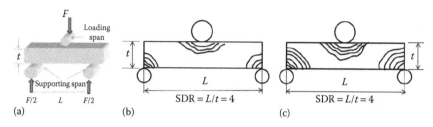

FIGURE 2.1

(a) Schematic of 3-point bend test; and stress distribution in a (b) thin and (c) thick sample. (From Kim, J.-K. and Mai, Y.-W., *Engineered Interfaces in Fiber Reinforced Composites*, Elsevier, Amsterdam, the Netherlands, 1998.)

induced in a beam subjected to a bending load is directly proportional to the magnitude of the applied load and independent of the span length. Thus, the support span of the short beam shear specimen is kept short so that an interlaminar shear failure occurs before a bending failure.

This test method is defined by ASTM D2344 [3], which specifies a span length to specimen thickness ratio of five for low stiffness composites and four for higher stiffness composites. This test has an inherent problem associated with the stress concentration and the nonlinear plastic deformation induced by the loading nose of small diameter. This is schematically illustrated in Figure 2.1, where the effects of stress concentration in a thin specimen are compared with those in a thick specimen.

Both specimens have the same span-to-depth ratio. The stress state is much more complex than the pure shear stress state predicted by the simple beam theory. According to the classical beam theory, the shear stress distribution along the thickness of the specimen is a parabolic function that is symmetrical about the neutral axis, where it is at its maximum and decreases toward zero at the compressive and tensile faces. In reality, however, the stress field is dominated by the stress concentration near the loading nose, which completely destroys the parabolic shear distribution used to calculate the apparent interlaminar shear strength (ILSS). Recognizing the deficiencies, many investigators have performed detailed studies of this test method.

2.2.3 Flexural Test

The flexural methods are applicable to polymeric composite materials. A testing machine with controllable crosshead speed is used in conjunction with a loading fixture. Theoretical estimation of the strength and modulus of laminated composites in tensile and compressive loading is quite simple and well established. Still, due to complex states of stress during flexural loading, the prediction of the mechanical properties becomes quite tedious and requires a great deal of material properties.

The load (P) ~ displacement (h) curve obtained from flexural testing was converted into stress (σ) ~ strain plots (ε) by the following relationships.

$$\sigma = \frac{3Pl}{2bd^2}$$

$$\varepsilon = \frac{6hd}{l^2}$$

l, b, and d represent the span length, width, and thickness of the sample, respectively. Intuitively, the strength of the material is the stress, which corresponds to the maximum load. The modulus has been determined from the slope of the initial linear region of the stress ~ strain curve.

2.3 Dynamic Mechanical Analysis

Dynamic Mechanical Thermal Analyzer (DMTA) is a tool to evaluate the viscoelastic response of the material for a wide range of temperatures. The instrument applies a dynamic load to the sample, and the response of the material is recorded in the form of dynamic displacement. For a perfectly elastic solid, the applied stress and resulted strain remain in phase, whereas there is a phase difference in the case of polymeric (viscoelastic) material. The storage modulus (E') obtained from DMTA is a representation of the elastic modulus of the material, whereas the loss modulus (E'') reflects the viscous modulus. The damping tendency of the material is determined from the parameter tan δ (ratio of E'' to E'). The E', E'', and tan δ are determined using the following equations.

$$E' = \frac{\sigma_o}{\varepsilon_o} \cos \delta \qquad\qquad (2.1)$$

$$E'' = \frac{\sigma_o}{\varepsilon_o} \sin \delta \qquad\qquad (2.2)$$

$$\tan \delta = \frac{E''}{E'} \qquad\qquad (2.3)$$

2.4 Scanning Electron Microscope

Fractography has become synonymous with an investigation in which both the mechanics of cracking are identified and the influence of the environment and/or the internal structures of the component on the mechanics of fracture are determined. The examination of the fracture surface of a polymer is usually carried out sequentially using low power optical microscopy, high power optical microscopy, and either SEM applied directly to the surface or transmission electron microscopy of replicas taken from the surface. The use of replicas has declined in the recent years due to the improvement and increased availability of scanning electron microscopes. The scanning electron microscope, due to its very large depth of field, is ideally suited to the study of fractures in polymers and can focus on even the most fibrous fracture surfaces. Magnifications between 20× and 10,000× are possible, and it allows correlation of the detailed structures observed at high magnification with the coarser optically visible features of the fracture.

To understand the interfacial bonding condition between the fiber and the matrix and its effect on mechanical properties, photomicrographs were taken using a scanning electron microscope. Dramatic changes in the structure and properties of the composite, when exposed to cryogenic temperatures, particularly in cyclical fashion, can be seen by studying the SEM micrographs. Generally, cryogenic cycling leads to microcracking, delamination, and potholing that are localized surface degradation [3]. Increased thermal stresses are the underlying cause of microcracking in composites at cryogenic conditions. As the temperature of the laminate falls below its stress-free temperature, residual stresses develop in the material. These stresses are the result of differences in coefficient of thermal expansion between the fibers and the matrix. As the temperature deviates from the stress-free temperature, the amount of the thermal stresses increase. And when these residual stresses become large enough, they are relieved through physical processes, such as microcracking, delamination, and potholing.

2.5 Atomic Force Microscopy

The AFM consists of a cantilever with a sharp tip (probe) at its end that is used to scan the specimen surface. The cantilever is typically silicon or silicon nitride with a tip radius of curvature on the order of nanometers. When the tip is brought into proximity of a sample surface, forces between the tip and the sample lead to a deflection of the cantilever according to Hooke's law.

AFM has become a useful tool for characterizing the topography and properties of solid materials since its advent. Besides topography information, the phase lag of the cantilever oscillation, relative to the signal sent to the cantilever's piezo driver, is simultaneously monitored giving information about the local mechanical properties, such as adhesion and viscoelasticity. Phase imaging is a powerful tool that provides nanometer scale information often not revealed by other microscopy techniques [4].

2.6 Differential Scanning Calorimetry Analysis

Differential scanning calorimetry (DSC) is a thermoanalytical technique in which the amount of heat required to increase the temperature of a sample and reference is measured as a function of temperature. Both the sample and reference are maintained at nearly the same temperature throughout the experiment.

Temperature modulated differential scanning calorimetry (TMDSC) is a modified thermoanalytical technique over a conventional one which involves superimposition of a modulation on the conventional linear heating, cooling, or isothermal temperature program. The temperature modulation can be described by sinusoidal function, linear sawtooth pattern, or some other mathematical expression.

A discrete Fourier transformation of the modulated heat flow (raw signal) yields the dynamic heat flow and deconvoluted (average) response. The latter is denoted as total heat flow. It can be separated into its heat capacity and kinetic components, also known as reversing and non-reversing heat flow, respectively. Another possibility of interpretation of the results is enabled by calculation of complex heat capacity, which represents the ratio of the modulated heat flow amplitude and modulated heating rate amplitude.

TMDSC is also known as Alternating DSC (ADSC). ADSC is based on the temperature modulation during a constant heating rate in the nonisothermal experiments, but the quasi-isothermal conditions mean that the temperature is altered in a sinusoidal fashion with an angular frequency ω (radian s^{-1}) and sufficiently small amplitude about a constant temperature.

2.7 Fourier Transform Infrared Spectroscopy Analysis

Infrared spectroscopy is a widely used technique for the identification of different chemical functional groups present in the composites. The infrared spectrum is divided in three regions. The normal region which is also known as mid infrared region extends from 4,000 to 667 cm^{-1} and results from vibrational plus rotational transitions. This region is particularly meant for organic chemists since the vibrations induced in organic molecules are absorbed in this region. The regions on either side of the infrared are known as near infrared (12,500 to 4,000 cm^{-1}) and far infrared (667 to 50 cm^{-1}) regions. The near infrared region exhibits bands assignable to harmonic overtones of fundamental bands and combination bands, whereas the far infrared region deals with the pure rotational motion of the molecules. The combination of infrared spectroscopy with the theories of reflection has made advances in surface analysis possible. The physical property that is measured in infrared spectroscopy is the ability of some molecules to absorb infrared radiation. Atoms in molecules are not static, as one might think, but rather they vibrate about their equilibrium positions. The frequency of these vibrations depends on the mass of the atom and the length and strength of the bonds. Molecular vibrations are stimulated by bonds absorbing radiation of the same frequency as their natural vibrational frequency (usually in the infrared region). FTIR works by exciting chemical bonds with infrared light and is best for identification of organic materials. The different chemical bonds in this excited state absorb the light energy at frequencies unique to the various bonds. This activity is represented as a spectrum. The spectrum can be expressed as % transmittance (%T) or % absorbance (%A) versus wavenumber. The wavenumber of the peak tells what types of bonds are present and the %T tells the signal strength. Low signal strength directly affects the resolution of the peaks, making sample size and preparation key for acquiring a quality spectrum [5]. The region of the infrared spectrum from 4,000 to 1,400 cm^{-1} exhibits absorption bands that fall under the functional group regions. These bands are useful diagnostically, but more usually they supplement the region below 1,400 cm^{-1}. The region from 1,400 to 900 cm^{-1} is complex because it contains, apart from fundamental stretching and bending vibrations, many bands resulting from the sum or difference of their vibration frequencies. Specific vibrational assignments in this region are, therefore, very difficult. Thus, this part of the spectrum is characteristic of a compound and is called the fingerprint region. Similar molecules may show very similar spectra in the functional group region, but certainly exhibit discernible differences in the fingerprint region. A ratio of specific peak heights can sometimes be used to quantify proportions in simple mixtures, degree of oxidation or decomposition, purity, and so on. The FTIR aids in identifying chemical bonds and, thus, chemical composition of materials.

References

1. A. Standard, Standard test method for tensile properties of polymer matrix composite materials, *ASTM D3039D 3039M*, 2008.
2. J.-K. Kim and Y.-W. Mai, *Engineered Interfaces in Fiber Reinforced Composites*. Amsterdam, the Netherlands: Elsevier, 1998.
3. A. Standard, Standard test method for short-beam strength of polymer matrix composite materials and their laminates, *Annu. Book ASTM Stand. West Conshohocken*, 15: 54–60, 2007.
4. J. Harding and Y. L. Li, Determination of interlaminar shear strength for glass/epoxy and carbon/epoxy laminates at impact rates of strain, *Compos. Sci. Technol.*, 45(2): 161–171, 1992.
5. B. C. Smith, *Fundamentals of Fourier Transform Infrared Spectroscopy*. Boca Raton, FL: CRC Press, 2011.

3

Temperature-Induced Degradations in Polymer Matrix Composites

Environmental temperature is one of the important parameters that decides the life and durability of the composite material in real-time application. The performance of the individual constituents and more importantly, the fiber/polymer interface at the in-service temperature (may be static or dynamic) is decisive in this regard. Hence, proper evaluation of the properties of the composite in its particular service temperature must be well ensured.

This is particularly the case for carbon fiber-reinforced polymer (CFRP) laminates used or of interest in aircraft and space structures, where the laminates may be exposed to air at −73°C to 80°C or the space at −140°C to 120°C [1].

The progressive changes that occur in the thermophysical and thermomechanical properties of FRP composites with increasing temperature result from the alteration in the molecular structure of their polymer components. The bonds existing in thermoset polymers (which have frequently been used as the resin in composite materials) can be divided into two major groups: (1) primary and (2) secondary. The first group includes the strong covalent intramolecular bonds in the polymer chains and cross-links. The dissociation energy of such bonds varies between 50 and 200 kcal/mol. Secondary bonds include much weaker bonds, for example, hydrogen bonds (dissociation energy: 3–7 kcal/mol), dipole interaction (1.5–3 kcal/mol), and Van der Waals interaction (0.5–2 kcal/mol). Consequently, secondary bonds can be more easily dissociated. When the temperature increases, secondary bonds are broken during glass transition and the material state changes from glassy to leathery. As the temperature is raised further, the polymer chains form entanglement points where molecules, because of their length and flexibility, become knotted together. This state, designated "the rubbery state," is also characterized by intact primary and broken secondary bonds, but in an entangled molecular structure. When even higher temperatures are reached, the primary bonds are also broken and the material decomposes, which is known as the decomposition process.

The effect of environmental temperature on the mechanical performance of a composite may be studied in two different ways: (1) *in-situ* temperature mechanical testing and (2) mechanical testing after a long- or short-term thermal conditioning.

3.1 *In-Situ* Temperature Mechanical Performance

This type of testing is of particular interest, when the material is subjected to an external load at above- and below-ambient temperatures during its service.

3.1.1 Elevated Temperature Mechanical Performance

FRP composites are sensitive to temperature variations as a result of induced thermal stresses between the fibers and the polymer matrix [2] due to their distinct thermal expansion coefficients. At elevated temperatures, differential thermal expansion of the fiber and the matrix may lead to the formation of microcracks at the fiber/polymer interface [3]. The fiber/matrix interface also becomes susceptible to aggressive reactions under the exposure of high temperature environments, which can lead to the degradation of both the fibers and the matrix [4]. This, in turn, affects the integrity of the composites, since it is the interface through which the thermal and mechanical loads are transferred from the matrix to the fibers.

In-situ mechanical testing for the evaluation of interlaminar shear strength (ILSS) at different temperatures demonstrated that the environmental temperature has a profound effect on the properties, which are primarily dependent on the fiber/matrix interfacial adhesion [5]. Figure 3.1 shows the effects of above ambient temperature environments on the ILSS of glass fiber/epoxy (GE), carbon fiber/epoxy (CE), and Kevlar fiber/epoxy (KE) composites.

The reduction in ILSS for the GE composites when tested at +50°C and +100°C was about 47% and 87%, respectively as compared to their room temperature (RT) values. When the CE composites were tested at +50°C and +100°C, the reduction in ILSS (as compared to their RT values) was about 81%

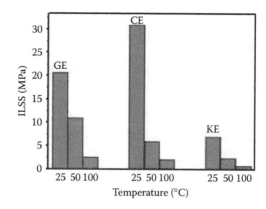

FIGURE 3.1
Effect of different *in-situ* elevated temperatures on ILSS of different FRP composites (GE: glass fiber/epoxy, CE: carbon fiber/epoxy and KE: Kevlar fiber/epoxy). (From Sethi, S. et al., *Mater. Des.*, 65, 617–626, 2015.)

and 93%, respectively. For the Kevlar fiber composites, this reduction in ILSS was 65% and 90%, respectively [5].

The effect of a thermal environment on the residual mechanical performance was evaluated for a constant 170°C temperature for 120, 240, and 626 h prior to flexural testing [6]. Both the flexural and shear strength decreased and became more pronounced at prolonged exposure times due to the weakening of the interface. It was also shown that the fracture changes with the increase in thermal ageing time from a ductile appearance containing large plastic deformation to a more brittle appearance, as shown in Figure 3.2.

The ILSS of unidirectional graphite composites were decreased by 30% under the exposure of elevated temperatures [7]. Investigations on CFRP and glass fiber-reinforced plastic (GFRP) sheets, which are exposed to 600°C, reported that the residual tensile strength and stiffness severely degraded when the composites are exposed to a temperature higher than the decomposition temperature of polymer resin and a further increase in environmental temperature would not lead to any further reduction in the aforesaid properties [8]. Keller et al. investigated the response of liquid-cooled GFRP slabs in the fire [9]. Wu et al. studied the tensile behavior of CFRP sheets with two different epoxy resins at different temperatures [10]. An investigation on epoxy resin at elevated temperature elucidated that the compressive strength of epoxy significantly decreased at temperatures close to the glass transition temperature and reached to zero with further increase in temperature [11].

The effects of extreme temperature variations were experimentally studied on impact damage to CFRP laminates. As the temperature of a CFRP laminate increases, the delamination areas of impact-induced damage decreases. Therefore, it is proven that temperature significantly influences the impact damages of CFRP laminates. In the case of extreme low/high temperatures, a linear relation between the impact energy and the delamination areas was observed [1].

In fact, low temperatures produce interlaminar residual thermal stresses, high enough to accelerate matrix cracking during low velocity impact, with consequent delamination of the composite laminate [12]. On the other hand,

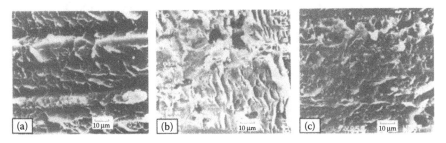

FIGURE 3.2
Fracture surfaces after dry thermal ageing for (a) 120 h, (b) 240 h, and (c) 626 h. (From Birger, S. et al., *Composites*, 20, 341–348, 1989.)

fiber-reinforcement architecture and stacking sequence play an essential role in the behavior of composite laminates under such thermal and loading conditions [13].

Río et al. [13] compared the low velocity impact response of unidirectional, cross-ply, quasi-isotropic, and woven carbon-epoxy laminates in low temperature conditions. The threshold energy (impact energy below for which no apparent damage is introduced within the laminate) decreased up to 50% in quasi-isotropic laminates for temperatures between 20°C and −150°C.

Icten et al. [14] studied the impact behavior of glass/epoxy laminates, with a stacking sequence of $[0/90/45/−45]_s$, at low temperatures. This study was performed for temperatures of 20°C, −20°C, and −60°C and impact energies that ranged from 5 to 70 J. The impact response and the damage tolerance of the composite was found to be nearly the same for all temperatures up to the $E_i = 20$ J, as shown in Figure 3.3. The primary damage mechanism up to this energy level seems to be matrix cracking, delaminations, and back surface delaminations. Beyond that energy, temperature affects significantly the impact behavior, where the fiber breakages and back surface delaminations become dominant damage modes. Finally, they observed that the perforation threshold increases with the decrease of test temperature [13].

Unidirectional and cross-ply glass-epoxy laminates were studied by Ibekwe et al. [15], where the temperatures ranged from 20°C to −20°C, and they concluded that more damages were induced with decreasing temperature.

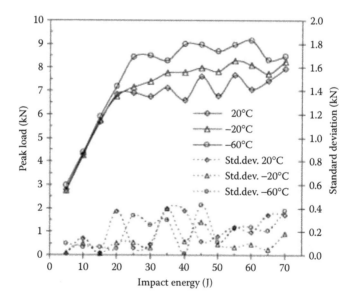

FIGURE 3.3
Variation of peak load with impact energy at different temperatures. (From Icten, B.M. et al., *Compos. Struct.*, 91, 318–323, 2009.)

3.1.2 Low and Cryogenic Mechanical Performance

Several investigations have been performed for the assessment of mechanical properties of FRP composites under cryogenic temperatures [16–19]. Under low temperature environments or cycling from RT to low temperature applying load, microcracks may generate and propagate in the polymer matrix and/or at the fiber/matrix interface [20]. Various structural damages such as fiber/matrix interfacial debonding, potholing, or delamination [18,19] result in the degradation of mechanical properties of FRP composites. Unidirectional CFRP laminate specimens that had been aged at −184°C for 555 h with half of the failure load showed about 20% degradation in tensile strength compared to that at RT [18]. The effects of low temperature and low temperature cycling (RT to −50°C, RT to −100°C, and RT to −150°C) on the strength and stiffness of T700/epoxy unidirectional laminates were reported [21], as shown in Figure 3.4a and b.

Figure 3.4a shows that there is an increase in stiffness as temperature decreased and at −150°C, the stiffness increased about 16% more than at RT. It can also be noted from the figure that the rate of increase in stiffness is higher between RT and −50°C and then gradually lessened below −50°C, and the strength of noncycled composite specimens decreased about 9% more at −150°C than at RT. SEM micrographs revealed the failure of the interface as it is clearly removed in the RT failed specimens while cryogenic temperature specimens and cryogenic temperature cycled specimens showed the presence of an interface (Figure 3.5). Low thermomechanical cycles can result in improvement in the interfacial shear strength of laminated composites [19].

Apparent ILSS variation with the testing temperature for GE, CE, and KE composites is shown in Figure 3.6.

The mechanical behavior of carbon/epoxy [22] and glass/epoxy [23,24] at liquid nitrogen temperature has been investigated. The roles of

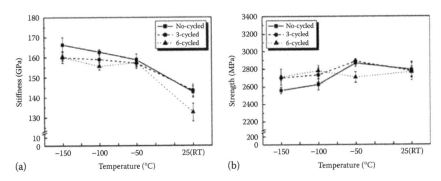

(a)   (b)

FIGURE 3.4
Results of stiffness and strength of T700/epoxy unidirectional specimen cycled with load at RT/−50°C, /−100°C, and /−150°C. (a) Change of stiffness and (b) change of strength. (From Kim, M.-G. et al., *Compos. Struct.*, 79, 84–89, 2007.)

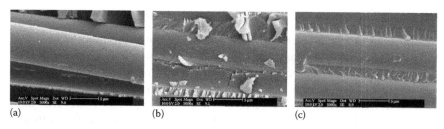

(a)   (b)   (c)

FIGURE 3.5
Interfaces of fractured T700/epoxy unidirectional specimens through SEM images. (a) at RT,
(b) at −150°C, and (c) at −150°C after six cycles from RT~−150°C. (From Kim, M.-G. et al.,
Compos. Struct., 79, 84–89, 2007.)

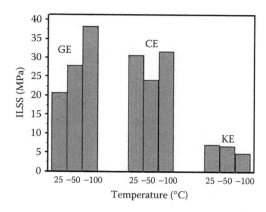

FIGURE 3.6
Effect of different *in-situ* low temperatures on ILSS of different FRP composites. (From Sethi, S.
et al., *Mater. Des.*, 65, 617–626, 2015.)

percentage matrix phase and interfacial areas on the interlaminar shear
failure mechanism of glass/epoxy composites at ultra-low temperatures
for different loading speeds were reported [23]. The stress at delamination failure was found to increase slightly with increased crosshead
speed at some points of conditionings for three temperatures. The ILSS
value increased with more conditioning time for the same loading rate.
Here, no statistically significant variations of shear strength values were
found with the change of conditioning temperature. Greater value of
shear strength at a longer conditioning time may possibly be attributed
by the enhanced key and lock principle at the fiber/polymer interfaces.
A greater percentage of fibers in the composites has resulted in more
interfacial areas and thus more matrix damages were induced by a misfit
strain (Figure 3.7).

Three weight percentages of woven carbon fibers (50, 55, and 60 wt%)
were utilized to prepare the CE laminates. The moisture-free carbon/epoxy

FIGURE 3.7
SEM micrograph reveals total loss of adhesion at the fiber/polymer interface and massive matrix cracking at low temperature. (From Ray, B.C. et al., *J. Appl. Polym. Sci.*, 100, 2289–2292, 2006.)

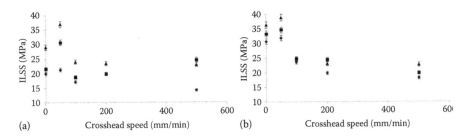

FIGURE 3.8
Graph showing the effect of crosshead speed on ILSS of carbon/epoxy composites at ambient temperature (triangle symbol), cryogenic temperature (77 K) (diamond symbol), and at room temperature after cryogenic conditioning (square symbol) for (a) 0.5 fiber weight fraction and (b) 0.6 fiber weight fraction. (From Surendra Kumar, M. et al., *J. Reinf. Plast. Compos.*, 28, 2013–2023, 2009.)

composite specimens were exposed to a liquid nitrogen environment (77 K) for 1 h (Figure 3.8) [22].

From the plots, it is evident that specimens tested at a cryogenic temperature show lower ILSS values than the untreated one. The cryogenic conditioning causes matrix hardening due to contraction, leading to stone-like structures in which disentanglement is almost absent. The anisotropic behavior of carbon fibers plays a critical role. SEM micrographs revealed fiber/matrix debonding and formation of shear cusps in cryogenically conditioned carbon/epoxy composites, as shown in Figure 3.9.

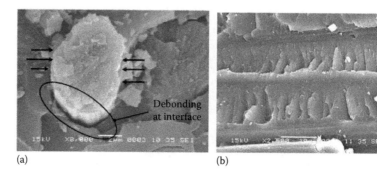

(a)   (b)

FIGURE 3.9
Scanning electron micrographs showing (a) debonding at interface and (b) rows of cusps. (From Surendra Kumar, M. et al., *J. Reinf. Plast. Compos.*, 28, 2013–2023, 2009.)

3.2 Effects of Thermal Cycling on Mechanical Behavior of Fiber-Reinforced Polymer Composites

3.2.1 Thermal Shock Cycling

When engineering and structural components undergo a sudden change in the temperature of their operative environment, then the materials are said to have undergone "thermal shock." Fiber/matrix interfacial adhesion is significantly affected by the thermal shock conditionings [24–28]. An investigation which was focused on the effect of thermal shock on the ILSS of glass/epoxy composite has reported that there is an increase in ILSS value with the increase in conditioning time, as shown in Figure 3.10 [25].

Here, the effect of thermal shock is evident at less conditioning time (at 5 min). But postcuring and strengthening phenomena and shrinkage compressive forces dominate at higher conditioning time. Shrinkage compressive

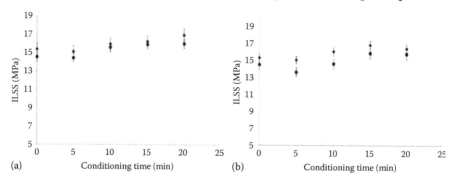

(a)   (b)

FIGURE 3.10
Effect of thermal shock on ILSS value of glass fiber/epoxy composites at 2 mm/min (sphere symbol) and 10 mm/min (diamond symbol) crosshead speeds (a) down-thermal cycle and (b) up-thermal cycle. (From Ray, B.C., *Mater. Lett.*, 58, 2175–2177, 2004.)

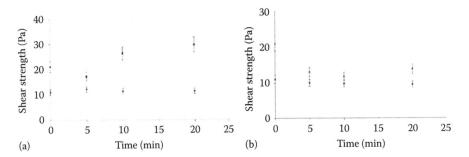

FIGURE 3.11
Effect of thermal shock on ILSS value of Kevlar fiber/epoxy (sphere symbol) and Kevlar fiber/polyester (diamond symbol) composites after (a) thermal and (b) cryogenic conditioning. (From Ray, B.C., *J. Mater. Sci. Lett.*, 22, 201–202, 2003.)

forces can improve the mechanical interlocking at the fiber/matrix interface. An investigation on glass fiber/polyester composite reveals that the degradation of the composite under hydrothermal shock cycle is less pronounced with the increase of the fiber volume fraction [26]. The damaging effects are also loading-rate sensitive. ILSS values were found higher for higher loading rates. The decrease in ILSS value with the increase in hydrothermal shock cycles was also noticed.

The influence of thermal shock on the mechanical behavior of Kevlar fiber reinforced with epoxy and polyester resin matrix was also studied [27]. For the Kevlar fiber/polyester system, which is thermally conditioned, the ILSS value decreases with increasing conditioning time, whereas the ILSS increases with conditioning time for Kevlar/epoxy composites, as shown in Figure 3.11. The weak interface between the Kevlar fiber and polyester resin deteriorates by thermal shock. The interface may be unable to accommodate the unfavorable tensile stresses developed due to the radial expansion of the Kevlar fiber.

3.2.2 Thermal Fatigue

When continuous fiber-reinforced composites are subjected to temperature variations, local stresses are generated in the composites due to the different coefficients of thermal expansion and/or due to the ply orientation in the lay-up [29]. When these thermal variations are cyclic, they result in cyclic stress variation at the ply level, which can be compared to a fatigue phenomenon. These cyclic stresses may result in various types of damage similar to those observed in mechanical fatigue like transverse matrix cracking, fiber/matrix debonding, and delamination. These temperature variations become more deleterious in the presence of an oxidative environment, which can cause matrix oxidation. Studies on oxidation phenomenon in thermoset epoxy polymers have reported that the oxidation results in a loss of mass and a reduction in volume of the epoxy matrix, inducing shrinkage of the matrix with respect to the fibers [29–31]. Oxidation of epoxy and bismaleimide may result in the

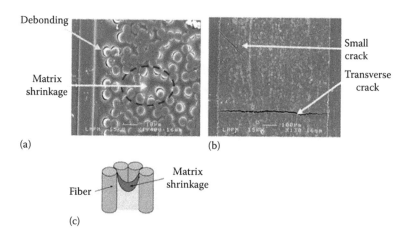

FIGURE 3.12

Different types of damages in $[0_3/90_3]$s laminate subjected to 100 cycles in oxygen: (a) fiber/matrix debonding, (b) transverse crack, and (c) permanent matrix deformation. (From Lafarie-Frenot, M.C. and Rouquie, S., *Compos. Sci. Technol.*, 64, 1725–1735, 2004.)

formation of a low thickness oxidized layer on the free edges of the samples, and during ageing, the thickness of this layer grows quickly toward an asymptotic value, governed by the type of material and temperature of the ageing atmosphere [32]. Throughout the thermal cycling tests, the observations of the free edges have shown three types of damages: (1) debonding between fibers and matrix (Figure 3.12a), (2) matrix cracking (Figure 3.12b), and (3) permanent deformation of the matrix due to its shrinkage (Figure 3.12c) [29].

The SEM picture of the edge of the quasi-isotropic laminate $[45/0/-45/90]$s subjected to 500 thermal cycles in oxygen revealed the presence of a deep hole in the matrix, which was located at the interface between the central 90° layer and a 45° ply. In that case, the matrix contraction has induced numerous fiber/matrix debondings in the circumference of the hole. When the tests have been performed in an oxidative environment, it appeared that, after 100 cycles, matrix shrinkage is already well developed. Consequently, fiber/matrix debonding, which seems to be connected with this type of damage, has been observed very early during the test. For example, in Figure 3.13a, some small fiber/matrix debondings are present in a $[0_3/90_3]$s specimen subjected to only 100 thermal cycles in air. Moreover, observations of the same specimen, at the same stage of the test, have shown short matrix cracks (Figure 3.13b).

It is shown that the onset of fiber-matrix debonding depends on the oxygen pressure, and the increase of oxygen pressure will accelerate the thermooxidation phenomena by decreasing the time of the fiber-matrix debonding onset. When thermal cycling is performed in an oxidative atmosphere, there exists a coupling between thermal transverse stresses and oxidation, which enhances the importance of the location of the layers with respect to the atmosphere [33]. A research program entitled COMEDI was aimed to better identify the link between the physical mechanisms involved in thermooxidation

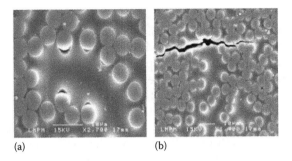

(a)   (b)

FIGURE 3.13
SEM observations of a $[0_3/90_3]$s specimen subjected to 100 thermal cycles in air: (a) interfacial debonding and (b) matrix micro-crack. (From Lafarie-Frenot, M.C. and Rouquie, S., *Compos. Sci. Technol.*, 64, 1725–1735, 2004.)

phenomena: chemical shrinkage strain/stress, oxygen reaction diffusion, degradation at different scales, and to provide tools for predicting the thermooxidation behavior of composite materials under thermooxidative environments including damage onset [34].

3.3 Effects of Fire Exposure on Fiber-Reinforced Polymer Composites

The transportation industry, which includes automobiles, marine craft, trains, and airplanes, is one of the largest consumers of composites. In transportation, the use of FRP has been primarily on interior surface finish and parts that require special shapes that are difficult/impossible to achieve with conventional materials (i.e., steel and aluminum). In recent years, FRP has been used more in structural applications. Some examples include railcar floor systems, naval ship bulkheads and decks, aircraft bodies, and reinforcement in structural columns. One of the primary design challenges that remains with FRP materials is their fire performance. The fire performance requirements depend on the material application and may include fire resistance, flammability, as well as smoke and toxic gas generation. The initial step in predicting the material behavior for all fire performance areas is to determine the heat and mass transfer through the material. For combustible materials such as FRP, this is done using a pyrolysis model that includes temperature-dependent thermal properties, decomposition kinetics and energetics, and a means to predict pyrolysis gas flow out of the material.

In most of the structural applications, such as pedestrian bridges or decks of road bridges and other civil constructions, FRP pultruded profiles are used, but still the full potential is hindered due to the susceptibility of the mechanical properties to elevated temperature and fire [35]. Modern building codes specify requirements for the fire resistance performance of construction elements used in different parts of a building. Structural elements are expected to afford fire resistance consistent with the building type/importance and geometry to prevent structural collapse under fire and allow the safe evacuation of occupants. As an example, the Portuguese fire safety code [36] specifies minimum fire resistance ratings of 30, 60, and 90 min for standard residential houses up to 9, 28, and 50 m tall, respectively. Under the Swiss code, the requirements are slightly stricter: 30 min for 2-story buildings, 60 min for 3-story buildings, or 90 min for taller buildings. In light of these typical fire resistance requirements, the experimental studies described in this section confirm the susceptibility of pultruded GFRP members to fire. The literature review also shows that such susceptibility is strongly dependent on: (1) the structural function of the GFRP members, (2) the type of fire exposure they are subjected to, (3) the cross-section, namely, the walls' thickness and geometric configuration (open, closed, or multicellular), and (4) the use of passive or active fire protection systems.

The tensile strength of pultruded GFRP composites at RT is dominated by their fibers. However, this might only be valid at elevated and high temperatures if the polymer resin can efficiently transfer the stresses between discontinuous fibers. Bai and Keller [37] and Correia et al. [38] tested GFRP specimens in tension up to failure at different temperatures ranging from 20°C to 220°C. Bai and Keller [37] observed typical tensile failure modes (i.e., broken fibers due to tension) up to 100°C; at higher temperatures, specimens failed in the clamp region because the polymer resin became soft and the entire roving layer was pulled out from the two outer mat layers (without any fiber breakage)—therefore, the results were reported up to 100°C. In the tests performed by Correia et al. [38], where grip failure was prevented by keeping the clamps at ambient temperature, failure occurred in the gage (heated) region due to the tensile rupture of the glass fibers for the entire temperature range. The normalized strengths presented in the results show that the reduction in tensile strength is much lower than that of shear and compressive strengths, as the former property is much less resin dominated.

Bai and Keller [37] and Correia et al. [38] also measured the in-plane shear strength of pultruded GFRP material at different temperatures from 20°C to 250°C (covering glass transition), using a 10 off-axis configuration. It appears that the shear strength degradation (very consistent in those two studies) is mainly dominated by the polymer resin, because nearly no fiber breakage was involved during the shear failure process. Correspondingly, the decrease of shear strength with temperature agrees well with that of the storage modulus curve measured from dynamic mechanical analysis tests.

Pultruded FRP composites show more complex behavior in compression when the temperature increases. This involves global buckling for slender specimens and local buckling or kink-band failure for compact specimens [39]. More importantly, the effective slenderness of identical specimens may change considerably with the increase of temperature because of the decrease in the material modulus. This also causes changes in failure modes. Compact pultruded GFRP specimens were tested in compression by Wang et al. [40] on C-channel sections, by Bai and Keller [37] on circular tube sections, and by Correia et al. [38] on I-sections. As shown in Figure 3.6, the measured material compressive strengths at different temperatures normalized by that at 20°C were consistent for those studies (and so were the failure modes), indicating a more significant degradation compared to shear and tensile strengths.

The studies described in this section do, in fact, draw attention to the importance of adopting an architectural design in which GFRP beams are embedded in floors and GFRP columns are integrated in the facades and/or embedded in the partition walls [38].

References

1. K.-H. Im, C.-S. Cha, S.-K. Kim, and I.-Y. Yang, Effects of temperature on impact damages in CFRP composite laminates, *Composites Part B*, 32(8): 669–682, 2001.
2. M. W. Hyer, C. T. Herakovich, S. M. Milkovich, and J. S. Short, Temperature dependence of mechanical and thermal expansion properties of T300/5208 graphite/epoxy, *Composites*, 14(3): 276–280, 1983.
3. D. L. Hiemstra and N. R. Sottos, Thermally induced interfacial microcracking in polymer matrix composites, *J. Compos. Mater.*, 27(10): 1030–1051, 1993.
4. H. Aglan, Z. Qian, and D. Mitra-Majumdar, The effect of temperature on the critical failure properties of advanced polymer composites, *Polym. Test.*, 11(3): 169–184, 1992.
5. S. Sethi, D. K. Rathore, and B. C. Ray, Effects of temperature and loading speed on interface-dominated strength in fibre/polymer composites: An evaluation for in-situ environment, *Mater. Des.*, 65: 617–626, 2015.
6. S. Birger, A. Moshonov, and S. Kenig, The effects of thermal and hygrothermal ageing on the failure mechanisms of graphite-fabric epoxy composites subjected to flexural loading, *Composites*, 20(4): 341–348, 1989.
7. C. E. Lundahl and J. H. Kreiner, Effect of composite properties on interlaminar shear strength, in *National SAMPE Symposium and Exhibition (Proceedings)*, pp. 1499–1503, 1986.
8. S. K. Foster and L. A. Bisby, High temperature residual properties of externally bonded FRP systems, in *Proceedings of the 7th International Symposium on Fiber Reinforced Polymer Reinforcement for Reinforced Concrete Structures (FRPRCS-7)*, SP-230-70, pp. 1235–1252, 2005.
9. T. Keller, C. Tracy, and A. Zhou, Structural response of liquid-cooled GFRP slabs subjected to fire–Part II: Thermo-chemical and thermo-mechanical modeling, *Composites Part A*, 37(9): 1296–1308, 2006.
10. Z. Wu, K. Iwashita, S. Yagashiro, T. Ishikawa, and Y. Hamaguchi, Temperature dependency of tensile behavior of CFRP sheets, *JSCMJ*, 32(3): 137–144, 2006.

11. J. M. Plecnik, R. Iding, J. D. Cunningham, and B. Bresler, Temperature effects on epoxy adhesives, *J. Struct. Div.*, 106(1): 99–113.
12. A. M. Amaro, P. N. B. Reis, and M. A. Neto, Experimental study of temperature effects on composite laminates subjected to multi-impacts, *Composites Part B*, 98: 23–29, 2016.
13. T. Gómez-del Río, R. Zaera, E. Barbero, and C. Navarro, Damage in CFRPs due to low velocity impact at low temperature, *Composites Part B*, 36(1): 41–50, 2005.
14. B. M. Icten, C. Atas, M. Aktas, and R. Karakuzu, Low temperature effect on impact response of quasi-isotropic glass/epoxy laminated plates, *Compos. Struct.*, 91(3): 318–323, 2009.
15. S. I. Ibekwe, P. F. Mensah, G. Li, S.-S. Pang, and M. A. Stubblefield, Impact and post impact response of laminated beams at low temperatures, *Compos. Struct.*, 79(1): 12–17, 2007.
16. J. B. Schutz, Properties of composite materials for cryogenic applications, *Cryogenics*, 38(1): 3–12, 1998.
17. P. K. Dutta and D. Hui, Low-temperature and freeze-thaw durability of thick composites, *Composites Part B*, 27(3): 371–379, 1996.
18. V. T. Bechel, J. D. Camping, and R. Y. Kim, Cryogenic/elevated temperature cycling induced leakage paths in PMCs, *Composites Part B*, 36(2): 171–182, 2005.
19. J. F. Timmerman, M. S. Tillman, B. S. Hayes, and J. C. Seferis, Matrix and fiber influences on the cryogenic microcracking of carbon fiber/epoxy composites, *Composites Part A*, 33(3): 323–329, 2002.
20. S. Sethi and B. C. Ray, Mechanical behavior of polymer composites at cryogenic temperatures, in *Polymers at Cryogenic Temperatures*, S. Kalia and S.-Y. Fu, (Eds.). Berlin, Germany: Springer, pp. 59–113, 2013.
21. M.-G. Kim, S.-G. Kang, C.-G. Kim, and C.-W. Kong, Tensile response of graphite/epoxy composites at low temperatures, *Compos. Struct.*, 79(1): 84–89, 2007.
22. M. Surendra Kumar, N. Sharma, and B. C. Ray, Microstructural and mechanical aspects of carbon/epoxy composites at liquid nitrogen temperature, *J. Reinf. Plast. Compos.*, 28(16): 2013–2023, 2009.
23. M. Surendra Kumar, N. Sharma, and B. C. Ray, Mechanical behavior of glass/epoxy composites at liquid nitrogen temperature, *J. Reinf. Plast. Compos.*, 27(9): 937–944, 2008.
24. B. C. Ray, Loading rate effects on mechanical properties of polymer composites at ultralow temperatures, *J. Appl. Polym. Sci.*, 100(3): 2289–2292, 2006.
25. B. C. Ray, Thermal shock on interfacial adhesion of thermally conditioned glass fiber/epoxy composites, *Mater. Lett.*, 58(16): 2175–2177, 2004.
26. B. C. Ray, Effect of hydrothermal shock cycles on shear strength of glass fiber-polyester composites, *J. Reinf. Plast. Compos.*, 24(12): 1335–1340, 2005.
27. B. C. Ray, Study of the influence of thermal shock on interfacial damage in thermosetting matrix aramid fiber composites, *J. Mater. Sci. Lett.*, 22(3): 201–202, 2003.
28. B. C. Ray, Thermal shock and thermal fatigue on delamination of glass-fiber-reinforced polymeric composites, *J. Reinf. Plast. Compos.*, 24(1): 111–116, 2005.
29. M. C. Lafarie-Frenot and S. Rouquie, Influence of oxidative environments on damage in c/epoxy laminates subjected to thermal cycling, *Compos. Sci. Technol.*, 64(10–11): 1725–1735, 2004.
30. S.-H. Lee, J.-D. Nam, K. Ahn, K.-M. Chung, and J. C. Seferis, Thermo-oxidative stability of high performance composites under thermal cycling conditions, *J. Compos. Mater.*, 35(5): 433–454, 2001.

31. M. S. Madhukar, K. J. Bowles, and D. S. Papadopoulos, Thermo-oxidative stability and fiber surface modification effects on the inplane shear properties of graphite/PMR-15 composites, *J. Compos. Mater.*, 31(6): 596–618, 1997.

32. A. Lowe, B. Fox, and V. Otieno-Alego, Interfacial ageing of high temperature carbon/bismaleimide composites, *Composites Part A*, 33(10): 1289–1292, 2002.

33. M. C. Lafarie-Frenot and N. Q. Ho, Influence of free edge intralaminar stresses on damage process in CFRP laminates under thermal cycling conditions, *Compos. Sci. Technol.*, 66(10): 1354–1365, 2006.

34. M. C. Lafarie-Frenot et al., Thermo-oxidation behaviour of composite materials at high temperatures: A review of research activities carried out within the COMEDI program, *Polym. Degrad. Stab.*, 95(6): 965–974, 2010.

35. J. R. Correia, Y. Bai, and T. Keller, A review of the fire behaviour of pultruded GFRP structural profiles for civil engineering applications, *Compos. Struct.*, 127: 267–287, 2015.

36. DL nº 220/2008, of 12 of November [Online]. Available: http://www.pgdlisboa. pt/leis/lei_mostra_articulado.php?nid=1949&tabela=leis. [Accessed: April 4, 2017].

37. Y. U. Bai and T. Keller, Modeling of strength degradation for fiber-reinforced polymer composites in fire, *J. Compos. Mater.*, 43(21): 2371–2385, 2009.

38. J. R. Correia, M. M. Gomes, J. M. Pires, and F. A. Branco, Mechanical behaviour of pultruded glass fibre reinforced polymer composites at elevated temperature: Experiments and model assessment, *Compos. Struct.*, 98: 303–313, 2013.

39. Y. Bai and T. Keller, Delamination and kink-band failure of pultruded GFRP laminates under elevated temperatures and compression, *Compos. Struct.*, 93(2): 843–849, 2011.

40. P. M. H. Wong, J. M. Davies, and Y. C. Wang, An experimental and numerical study of the behaviour of glass fibre reinforced plastics (GRP) short columns at elevated temperatures, *Compos. Struct.*, 63(1): 33–43, 2004.

4

Moisture-Dominated Failure in Polymer Matrix Composites

4.1 Background

In the last few decades, fiber-reinforced polymer (FRP) composites have gained significant attention as a structural material in marine environments. Corrosion problems associated with metallic structures and environmental degradation of wood have driven the process of replacing these materials fully or partially by FRP composites in naval applications. However, other advantages of FRP composites over metallic counterparts include the reduction of weight, particularly the topside weight of ships. Superior acoustic transparency of these composites has also directed their use in rodomes on ships and sonar domes on submarines. The noteworthy demand of FRP patrol boats is mostly due to their excellent resistance against corrosion, which reduces maintenance costs significantly and makes them light weight. Hence, better speed is achieved along with fuel economy. Offshore oil platforms are also a potential thrust application area of the glass FRP (GFRP) composite. In addition, GFRP is also being used in oil and water pipelines, water storage tanks, desalinization plants, tidal and wind turbines in the sea, and so on [1]. There also exist widespread civil engineering applications where FRP composites are being used, such as bridges, pedestrian footbridges, or repairing and strengthening of bridge concrete structures [2,3]. However, use of carbon FRP (CFRP) has been popular for these applications because of the low flexural modulus of GFRP that shows higher deflection in composite beam applications. Nevertheless, GFRP costs are much lower than CFRP, which drives the use of this material in applications where the high deflections and low flexural modulus of GFRP components do not hamper the application (Figure 4.1).

But, unfortunately, polymeric composites are susceptible to moisture. They absorb moisture in humid environments and undergo dilatational expansion. The presence of moisture and the stresses associated with moisture-induced expansion may cause lowered damage tolerance and structural durability. The structural integrity and lifetime performance of fibrous polymeric

FIGURE 4.1
Applications of FRP composites in (a) fishing boat, (b) ship, (c) stop log, (d) pipeline, and (e) wind turbine.

composites are strongly dependent on the stability of the fiber/polymer interfacial region. Moisture may penetrate into polymeric composite materials by diffusive and/or capillary processes [4,5]. The interactions between the fiber and the matrix resin are important rather complex phenomena. Both reversible and irreversible changes in mechanical properties of thermoset polymers are known to occur as a result of water absorption. This chapter illustrates various mechanisms of moisture diffusion in polymer matrix composites (PMC) and changes in the mechanical performance of these materials as a consequence upon moisture uptake.

4.2 Theories and Models of Moisture Uptake Kinetics

The moisture uptake rate and moisture content in the materials are strong functions of the diffusion temperature and water concentration (relative humidity or RH) in the environment. However, the material chemistry and water diffusion mechanism also affect the aforesaid water uptake behavior significantly [6]. Heterogeneity in the FRP composite makes the water uptake study quite complex. The moisture ingression pattern, moisture distribution in the material, and its consequences on the local and bulk performance of the FRP composites are some of the challenging tasks in this regard. Although, extensive work in this field has disclosed many underlying phenomena of water diffusion in FRP composites, still more concepts are yet to be revealed for a complete understanding of water interaction with FRP composites. Fick's diffusion model has been used frequently for the kinetic study of water diffusion in FRP composites. However, certain evidences have reported non-Fickian water uptake behavior of the FRP composite.

4.2.1 Fick's Model

Adolph Fick developed the simplest model in 1855, which is applicable to most polymeric composites [6], basing his work on the foundation set by Joseph Fourier [7]. Fick's first law of diffusion is based on the hypothesis that for an isotropic medium, rate of diffusion through any cross-section is directly proportional to the concentration gradient normal to it and is quantitatively represented as:

$$F = -D\left(\frac{\partial C}{\partial x}\right). \tag{4.1}$$

However, Fick's second law is considered as the fundamental law of diffusion and can be represented by Equation 4.2, when D is dependent on moisture concentration.

$$\frac{\partial C}{\partial t} = \frac{\partial(D\partial C)}{\partial x(\partial x)}. \tag{4.2}$$

However, in case of moisture concentration independent of D, Equation 4.2 becomes:

$$\frac{\partial C}{\partial t} = D\frac{\partial^2 C}{\partial x^2} \tag{4.3}$$

In the case of a water uptake study of such FRP composites, these equations have been simplified further. The water content (M) at any time (t) is determined by gravimetric analysis given by:

$$M = \frac{\text{weight of moist material} - \text{weight of dry material}}{\text{weight of dry material}}. \tag{4.4}$$

If diffusion is carried out at a constant temperature and relative humidity, the water content (M) in a thin flat plate kind of sample is given by:

$$M = G(M_\infty - M_i) + M_i. \tag{4.5}$$

This expression can be used during both absorption and desorption. M indicates moisture content, and the subscripts i and ∞ represent initial and at saturation stages, respectively. M_∞ is the maximum water content for that specific combination of temperature and relative humidity, respectively. G is a time-dependent parameter:

$$G = 1 - \frac{8}{\pi^2}\sum_{j=1}^{\infty}\frac{\exp\left[-(2j+1)^2\pi^2\left(\frac{D_z t}{s^2}\right)\right]}{(2j+1)^2}. \tag{4.6}$$

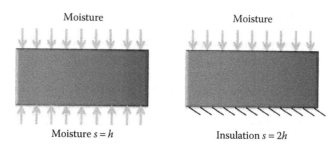

FIGURE 4.2

A schematic representation of diffusion process. (From Shen, C.-H. and Springer, G.S., *J. Compos. Mater.*, 10, 2–20, 1976.)

D_z represents diffusivity in the direction perpendicular to the plane of the plate, and s is related to the thickness of the plate (h) and is dependent on the available planes for the diffusion event, that is, if diffusion is allowed on both the parallel faces of the plate, then $s = h$, and if one face is insulated (water impermeable coating), then $s = 2h$, as shown in Figure 4.2.

However, attempts have been made to further simplify Equation 4.6 and approximated (Figure 4.3) by Shen and Springer [7]:

$$G = 1 - \exp\left[-7.3\left(\frac{D_z t}{s^2}\right)^{0.75}\right].$$ (4.7)

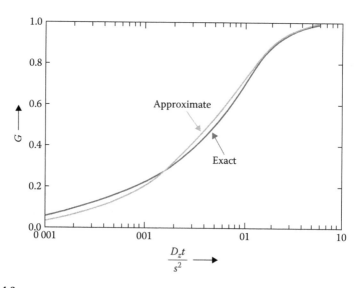

FIGURE 4.3

A comparison between exact and approximate solution. (From Shen, C.-H. and Springer, G.S., *J. Compos. Mater.*, 10, 2–20, 1976.)

For a unidirectional FRP composite, where the fibers are oriented at angles α, β, and γ with x, y, and z directions, respectively, the diffusion coefficient can be determined by

$$D_z = D_{11}\cos^2\gamma + D_{22}\sin^2\gamma. \tag{4.8}$$

D_{11} and D_{22} represent the diffusivity parallel (longitudinal) and perpendicular (transverse) to the fiber axis, respectively. However, it has been a common practice to determine the diffusion coefficient by plotting the parameter M versus $\sqrt{t}$, as shown in Figure 4.4.

In the earlier stage of moisture diffusion, the kinetics follow a linear relationship between M and $\sqrt{t}$, as the diffusion coefficient is independent of water content. The slope of this linear portion of the curve carries the information regarding the diffusivity, as per the expression given below.

$$\text{slope} = \frac{M_2 - M_1}{\sqrt{t_2} - \sqrt{t_1}} = \frac{4M_\infty}{h\sqrt{\pi}}\sqrt{D_{\text{eff}}}. \tag{4.9}$$

In the case where moisture diffusion does not take place through the edges of the sample or for an infinitely large flat sample, the diffusivity (D_z) becomes equal to D_{eff}. However, considering the edge effects, the diffusivity can be determined by taking the sample dimensions, that is, length (L), width (w) thickness into account.

$$D_z = D_{\text{eff}}\left(1 + \frac{h}{L} + \frac{h}{w}\right)^{-2}. \tag{4.10}$$

Multiple literature have reported Fickian diffusion in the FRP composites [8–13]. Fickian behavior is reported to be more pronounced when the polymer

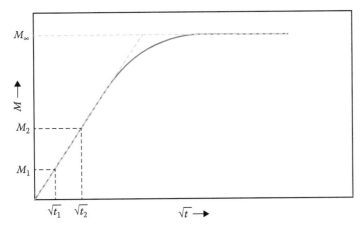

FIGURE 4.4
A representative moisture uptake kinetic curve.

composites are exposed to humid air and at lower temperatures. Numerous moisture absorption studies in polymer composites observed that the moisture uptake process follows the Fickian diffusion process, particularly in the initial linear uptake region [14–16]. Many other experimental investigations, however, found varying degrees of a non-Fickian moisture diffusion process [17–21]. Furthermore, the "Langmuir-type" model of diffusion [22], a time-varying diffusion coefficient model [20], are also presented. A study reveals that Langmuir and Fickian models could be statistically equivalent in some treatments and conditioning temperatures [23]. However, the Langmuir model should be favored when studying adhesive or carbon/epoxy conditioned in an anti-icing additive: it is a more accurate model and may capture the moisture uptake.

4.2.2 Langmuirian Diffusion Model

The Langmuirian model, also known as the dual-mode sorption model or the two-phase diffusion model, is based on the assumption that the penetrant molecules are divided into two populations. The first one is the preferential absorption of the diffused water molecules into the mobile phase of the material with diffusivity D_γ. Subsequently, the water molecules are constrained to the polymeric chains of the resin with a probability γ and the probability of becoming unbound is β. With these assumptions, Carter and Kibler [22] devised a modified Langmuirian model for analyzing the moisture absorption behavior in materials with non-Fickian diffusion characteristics. The approximated model is expressed below [24,25].

$$
M = M_\infty \left[\frac{\beta}{\beta+\gamma} \exp(-\gamma t) \left(1 - \frac{8}{\pi^2} \sum_{i=1}^{\infty(\text{odd})} \frac{\exp(-ki^2 t)}{i^2} \right) \right.
$$
$$
\left. + \frac{\beta}{\beta+\gamma} \left(\exp(-\beta t) - \exp(-\gamma t) + (1 - \exp(-\beta t) \right) \right]; 2\gamma, 2\beta \ll k \text{ and } k = \frac{\pi^2 D_\gamma}{h^2}.
$$

(4.11)

For shorter exposure timing, the above equation can be modified as

$$
M \approx M_\infty \frac{4}{\pi^{3/2}} \left(\frac{\beta}{\beta+\gamma} \right) \sqrt{kt} ; 2\gamma, 2\beta \ll k; t \le 0.7k.
$$

(4.12)

The same equation may be approximated for longer exposure time as follows:

$$
M \approx M_\infty \left(1 - \frac{\gamma}{\beta+\gamma} \exp(-\beta t); t \ge \frac{1}{k} \right)
$$

(4.13)

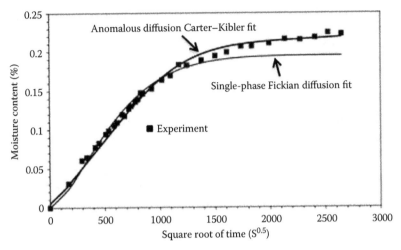

FIGURE 4.5
Fick's model and Langmuirian diffusion model applied to glass/epoxy composite exposed to humid ageing at 70°C and 85% RH. (From Kumosa, L. et al., *Compos. Part A Appl. Sci. Manuf.*, 35, 1049–1063, 2004.)

This model of two-phase diffusion has been adopted by many researchers to explain the moisture absorption kinetics of FRP composites [14,15,26–30]. However, the Langmuirian model has been reported to be able to accurately predict the moisture uptake of certain systems in certain environments, such as carbon/epoxy conditioned in an anti-icing additive.

As reported by Kumosa et al. [24], the modified Langmuirian model (i.e., Carter–Kibler model) fits more precisely to the experimental data than Fick's model, as shown in Figure 4.5.

4.2.3 Hindered Diffusion Model

Considering a one-dimensional (1D) diffusion process, at any point z and time t, the Langmuirian diffusion follows the below mentioned partial differential equations [21].

$$D\frac{\partial^2 n}{\partial z^2} = \frac{\partial n}{\partial t} + \frac{\partial N}{\partial t} \tag{4.14}$$

$$\frac{\partial N}{\partial t} = \gamma n - \beta N \tag{4.15}$$

The number of mobile and bound molecules per unit volume are represented by n and N, respectively. This process of absorption continues until the number of bound molecules, which are becoming mobile, becomes equal to the corresponding number of mobile molecules, which are becoming bound per unit time. Thus, the equilibrium is established when the following condition is met.

$$\gamma n = \beta N. \tag{4.16}$$

The water intake in an initial dry slab, which is exposed to environmental moisture at both of its faces, is given by

$$M = M_\infty \left[1 - \frac{8}{\pi^2} \sum_{i=1}^{\infty(\text{odd})} \left(\frac{r_i^+ \exp\left(-r_i^- t\right) - r_i^- \exp\left(-r_i^+ t\right)}{i^2 \left(r_i^+ - r_i^-\right)} \right) \right.$$

$$\left. + \frac{8}{\pi^2} \left(\frac{k\beta}{\beta + \gamma} \right) \sum_{i=1}^{\infty(\text{odd})} \left(\frac{\exp\left(-r_i^- t\right) - \exp\left(-r_i^+ t\right)}{r_i^+ - r_i^-} \right) \right] \tag{4.17}$$

where:

$$r_i^\pm = \frac{1}{2} \left[\left(ki^2 + \beta + \gamma\right) \pm \sqrt{\left(ki^2 + \beta + \gamma\right)^2 - 4k\beta i^2} \right].$$

This 1D hindered diffusion model (HDM) can be extended into 3D taking: (1) diffusion through all directions simultaneously and (2) interaction of diffusing molecules with each other into account.

$$D_x^* \frac{\partial^2 n^*}{\partial \left(x^*\right)^2} + D_y^* \frac{\partial^2 n^*}{\partial \left(y^*\right)^2} + D_z^* \frac{\partial^2 n^*}{\partial \left(z^*\right)^2} = \mu \frac{\partial n^*}{\partial t^*} + \left(1 - \mu\right)\left(n^* - N^*\right) \tag{4.18}$$

where:

$$n^* = \frac{n}{n_\infty} ; N^* = \frac{N}{N_\infty} ; t^* = \beta t$$

$$x^* = \frac{x}{l} ; y^* = \frac{y}{w} ; z^* = \frac{z}{h}$$

$$D_x^* = \frac{D_x}{l^2\left(\beta + \gamma\right)} ; D_y^* = \frac{D_y}{w^2\left(\beta + \gamma\right)} ; D_z^* = \frac{D_z}{h^2\left(\beta + \gamma\right)} ; \mu = \frac{\beta}{\beta + \gamma}$$

The dimensionless hindrance coefficient μ takes care of the temporal and spatial water concentration. It is also an indicative of the dependence of β and γ with the diffusion process. μ may have a value in between 0 and 1, depending on the rate of concentration change of the mobile water molecules. μ also governs the relationship between n^* and N^*, and, at equilibrium, both of these parameters approach unity. In case the hindrance coefficient becomes 1, the 3D HDM expressed in Equation 4.19 becomes the same as the 3D Fickian diffusion model as mentioned below:

$$D_x^* \frac{\partial^2 n^*}{\partial \left(x^*\right)^2} + D_y^* \frac{\partial^2 n^*}{\partial \left(y^*\right)^2} + D_z^* \frac{\partial^2 n^*}{\partial \left(z^*\right)^2} = \mu \frac{\partial n^*}{\partial t^*}. \tag{4.19}$$

It is of practical importance to determine the equilibrium moisture content in a polymer composite with a given set of diffusion parameters. Conventionally, this is confirmed from the experimental saturation value. However, defining saturation is crucial. American Society for Testing Materials (ASTM) D5229 suggests the attainment of saturation, when three consecutive gravimetric analyses don't show a difference of more than 0.01%. In case of polymeric composites, another important term is known as "pseudoequilibrium," which refers to the dramatic change in the rate of moisture absorption, as shown in Figure 4.6. Both Fickian and non-Fickian absorption models yield a linear relationship between M and $\sqrt{t}$ in the initial stage of diffusion, that is, before attainment of pseudoequilibrium.

However, the difference between Fickian and non-Fickian models is observed after this pseudoequilibrium stage. For the Fickian model, the

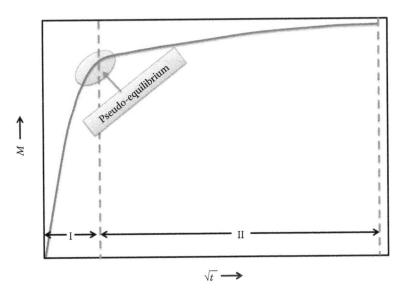

FIGURE 4.6
Schematic representation of both sections of hindered diffusion.

moisture content does not increase in the later stage of diffusion, suggesting that the equilibrium is competent to be the same as pseudoequilibrium. In case of the HDM, the moisture content still continues to increase after pseudoequilibrium, but at a very low rate. Hence, practically speaking, it is a very difficult and time-consuming process to predict the equilibrium content precisely.

Tang et al. [31] have reported the water uptake kinetics behavior of a CFRP sample in distilled water. They have compared the experimental data with a Fickian diffusion model. In the initial stage of water absorption, a perfect fit could be noticed. However, the experimental data at longer exposure time deviate from the Fickian model. Grace et al. [21] have fitted the same experimental data with an HDM and interestingly a nice fit could be obtained even at a longer time period of diffusion, as can be seen from Figure 4.7.

But the shorter experimental time frame limits the accurate and precise prediction of the equilibrium moisture content, that is, M_∞. In this case, a range of M_∞ values gives equally good fitting. The lowest value of M_∞ (1.85%) corresponds to a saturation time of 4.5 years, whereas the highest value (3.87%) indicates the saturation time to be 16 years. This implies that the short-term gravimetric water absorption analysis becomes insufficient to predict the accurate water content at equilibrium. The slow uptake rate in the later stage of an HDM requires a sufficiently long time for a better degree of accuracy.

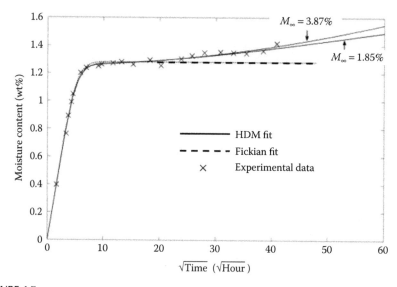

FIGURE 4.7
A comparison of Fickian and HDM fitting to experimental data. (From Grace, L.R. and Altan, M.C., *Compos. Part A Appl.+Sci. Manuf.*, 43, 1187–1196, 2012 and Tang, X. et al., *Compos. Sci. Technol.*, 65, 817–826, 2005.)

4.2.4 Dual-Stage Diffusion Model

A dual-stage diffusion model is based on the assumption that both Fickian and non-Fickian diffusion take place simultaneously during moisture diffusion in polymer composites. Different stages of diffusion are distinguished by the occurrence of the dominating diffusion mechanism, as shown in Figure 4.8.

In the beginning stage of diffusion, the Fickian model is dominating. With progress in time, the rate of relaxation increases and the diffusion rate slows down making the non-Fickian model as the dominating one [32]. If the moisture content due to Fickian and relaxation behavior are represented by M_F and M_R, respectively, the net water content (M) can be assumed to be the superpositioning of both of these behaviors [33]. This yields

$$M = M_F + M_R. \tag{4.20}$$

In this process, saturation will be attended when the relaxation process becomes completed. The Fickian contribution is simply rewritten from Equation 4.7, whereas the contribution of the relaxation process can be determined from the following expression:

$$M_R = M_{\infty,R}\left[1 - \exp(-kt)\right]. \tag{4.21}$$

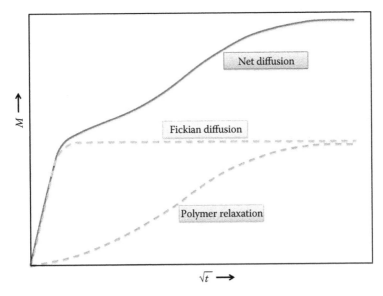

FIGURE 4.8
A representative dual-stage diffusion process.

where $M_{\infty,R}$ and k represent saturation moisture content for the relaxation mechanism and relaxation rate constant, respectively. Thus, Equation 4.20 becomes

$$M = M_{\infty,F}\left[1-\exp\left\{-7.3\left(\frac{Dt}{h^2}\right)^{0.75}\right\}\right]+M_{\infty,R}\left[1-\exp(-kt)\right]. \qquad (4.22)$$

Many articles have shown that the polymer composites follow dual-stage diffusion behavior. Jiang et al. [34] have reported a better fitting of experimental kinetic data with a dual-stage diffusion model than a Fickian model, as shown in Figure 4.9 for polyurethane adhesive. Fickian fitting at two different RH levels, but at the same temperature reveals no significant difference, which indicates that a Fickian model is less sensitive to RH difference.

It can also be observed from Figure 4.9 that an obvious deviation exists in the kinetic curves due to different RH levels after the linear Fickian stage. For water conditioning, moisture content is higher due to accelerated polymer relaxation. Kharbhari et al. [35] have also reported that dual-stage diffusion also fits well at various diffusion temperatures for a carbon/epoxy composite. As mentioned earlier, the initial linear portion indicates thermally activated Fickian diffusion, whereas at a later stage, a slow uptake rate can be observed. The later uptake mechanism can be derived from the synergetic effect of a number of mechanisms, which include polymer relaxation [36], filling of micropores or voids, and capillary action through the debonded interfaces [37] (Figure 4.10).

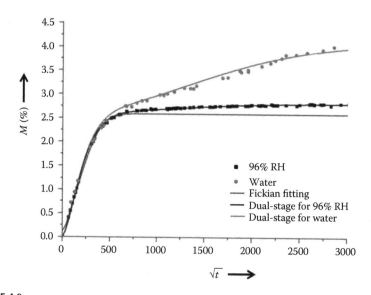

FIGURE 4.9
A comparison between Fickian and dual-stage fitting of experimental data at 40°C with different RH. (From Jiang, X. et al., *Compos. Part B Eng.*, 45, 407–416, 2013.)

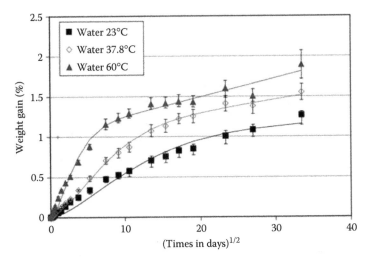

FIGURE 4.10

Water uptake curves for carbon/epoxy composites in deionized water at different temperatures. (Symbols represent experimental data and continuous curves represent corresponding dual-stage model fitting.) (From Karbhari, V.M. and Xian, G., *Compos. Part B Eng.*, 40, 41–49, 2009.)

4.3 Factors Affecting Moisture Uptake Kinetics in Fiber-Reinforced Polymer Composites

The rate of moisture intake and ultimate equilibrium moisture content in a FRP composite at a given combination of diffusion parameters to a large extent are dependent on the inherent moisture absorption tendency of the constituents (fiber and polymer) and subsequently generated interface.

4.3.1 Effect of Fiber

Affinity of the fiber toward environmental moisture is a matter of concern for designing and development of FRP composite with better moisture resistance. A schematic representation of a water diffusion path in FRP composites is given in Figure 4.11. Presence of a hydrophobic fiber (which is water impermeable) in a composite resists the diffusion of the water molecules through it, and, hence, forces the molecules to travel a tortuous extended path for its further diffusion into the bulk matrix (Figure 4.11b). This event thus delays the process of moisture diffusion to a significant extent. On the other hand, if the fiber used in the FRP composite is water permeable, diffusion of water molecules through this fiber becomes substantial (Figure 4.11a), and, thus, the FRP composite becomes less water resistance.

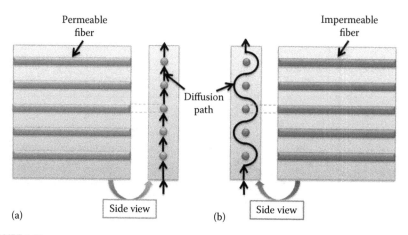

FIGURE 4.11

Schematic representation of the diffusion path of water molecules in FRP composites with (a) permeable fiber and (b) impermeable fiber.

Natural FRP composites are very susceptible to environmental moisture [38,39], as both the matrix and reinforcement are hydrophilic in nature. Presence of these kinds of permeable fibers is inefficient in retarding the moisture absorption tendency of the polymer. Glass and aramid fibers are partially water permeable in nature, and, hence, moisture absorption in polymeric composites containing these fibers is affected by both the fiber and resin. Moisture interaction with the metal oxides in E-glass leads to corrosion-induced damage and thus results in reduced mechanical strength [40,41]. On the contrary, carbon fibers are resistant to moisture absorption, and, therefore, moisture absorption in carbon FRP composites depends mostly on the matrix phase.

4.3.2 Effect of Polymer Matrix

The effect of a polymer matrix is much more prominent in deciding the moisture resistance of the FRP composite. Chemical structure, cross-link density, and crystallinity affect the moisture resistance of the polymer to a large extent. There exists a number of theories and models describing the moisture transport properties of polymers [42]. The "free volume" concept is one of these theories, which assumes that moisture penetrates into the polymer by means of the available free sites. In the amorphous or semi-crystalline polymer, it may be possible during curing that the polymeric chains are not closely packed and leave some nanopores, holes, or free volume with them. For epoxy, Soles and Yee [43] reported the average diameter of these nanopores varies from 5 to 6.1 Å, and they occupied 3%–7% of the total volume of this epoxy material. On the other hand, the approximate diameter of water molecule is 3.0 Å, and, hence, water molecule can easily traverse into epoxy through these nanopores [44]. The rate of diffusion and equilibrium moisture content thus are governed by the distribution and total volume of this free volume in the polymer. This free volume approach of explaining the moisture uptake behavior of a polymer is completely mechanistic and is erroneous in predicting the equilibrium moisture concentration. In addition, this model also fails to describe the diffusional behavior as a function of temperature and also the moisture plasticization effect in the polymer.

Another approach of describing the moisture uptake in polymers is the physicochemical interaction between the dispersant (moisture here) and host material (polymer). Apicella et al. [45] have proposed three modes of water diffusion into polymers: (1) water dissolution through the polymer chain network, (2) sorption into the free volume of the polymer, and (3) affinity of the polymer toward water molecules. The last point is referring to the hydrophilicity of a polymer and is related to the attractive forces between the polar polymeric group and water molecules, which is very often termed as hydrogen bonding [46]. The rate of water uptake, or diffusivity, is governed by the free volume of the polymer, whereas the equilibrium moisture content is decided by both free volume and number density of open hydrogen bonds in the polymer. Interestingly, some reports have also mentioned the influential role of polymer/water interaction on the diffusion coefficient [27,47].

Another approach of studying moisture uptake in polymers classifies the diffusing water molecules into two groups depending on their mobility [48] and specific interaction with the polymer [49]. The type-I water molecules are referred to as "free water molecules," which enter into the polymer network and break the interchain bonding like hydrogen and Van der Waals bonds and result in the swelling of the polymer. This essentially reduces the rigidity of the polymer due to the increased local segmental mobility of the chains and plasticization of the polymer. However, type-II (known as "bound") water molecules form multiple numbers of hydrogen bonds with the polymeric network, as shown in Figure 4.12. The water molecule thus bonded with many neighboring polymer chains forms an interconnected bridge structure, which acts as secondary cross-links and improves the stiffness of the polymer. A low temperature prefers the formation of type-I water molecules, whereas type-II bound water molecules are formed at relatively higher temperature and longer exposure time.

Apart from these mechanisms, the curing agent and its concentration, curing cycle, and method of composite fabrication also influence the moisture

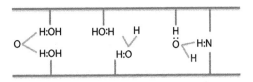

FIGURE 4.12
Formation of possible hydrogen bonding by type II water molecules.

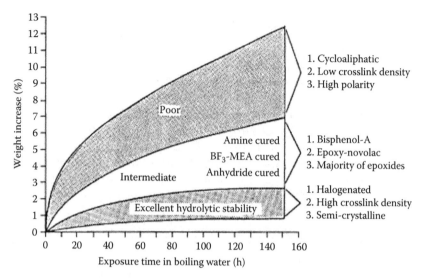

FIGURE 4.13
Effect of structural parameters of epoxy on water diffusion behavior. (Buehler, F.U. and Seferis, J.C., *Compos Part A Appl. Sci. Manuf.*, 31, 741–748, 2000 and Fleming, G.J. and Rose, T., Graphite composites for ocean ordinances, US Department of Commerce, National Technical Information Service Report AD-770-407, 1973.)

resistance of polymers [50]. A polymer with a higher cross-link density provides a better resistance toward moisture ingression [51]. Fleming and Rose [52] have provided the water absorption tendency of a variety of epoxy resins, as shown in Figure 4.13.

4.3.3 Effect of Interface

Fiber/matrix interface is believed to be an active pathway for the rapid diffusion of water into the polymer matrix composites. The diffusion can take place due to capillary action that is also known as "water wicking" and is predominant in the composite systems in which wetting of the fiber by the matrix is incomplete [53]. Like the polymeric phase, the fiber/matrix interface has also a strong impact on the water diffusion behavior of the FRP composite [54]. A good adhesion at the interface is the desirable criterion for having a good water resistance. Surface treatment of fibers is very often adopted to have a better interfacial bonding via electrostatic interaction or chemical bonding. Coupling agents in this context have been proven to be useful. A coupling agent usually exhibits two functional groups, which can react with both reinforcement and matrix, thus it integrates the system. Silane coating is usually provided on the surface of glass fibers, which acts not only as a protective coating, but also as a coupling agent to promote the adhesion with polymer matrix. The concept behind using silane coupling agents is to utilize chemical reactivity between the inorganic substrate and the organic resin, so as to develop proper adhesion at the fiber/matrix interface. A higher degree of cross-linking in the silane coupling layer results in a higher hydrolytic damage rate. Three reagents were used for silanization of glass fiber/epoxy composite, namely, 3-aminopropyltriethoxysilane (APTES), 3-aminopropylmethyldiethoxysilane, and 3-aminopropyldimethylmonoethoxysilane [55]. A composite, which had been fabricated from glass fibers coated with APTES, revealed the lowest diffusivity. The siloxane cross-linking structure of the APTES may be thought to have an impact on this diffusional behavior. The dense cross-linking structure of the interphase (of few tens of micrometers) yields a lower free volume for moisture retention and thus results in lower equilibrium water content.

4.3.4 Effect of Temperature

Temperature is likely to influence moisture pick-up kinetics in polymer composites in a complex manner [56]. The equilibrium moisture content in FRP composites is observed to be either independent of temperature or dependent on temperature [57]. Absorbed water is rarely distributed uniformly, and, thus, a distribution of internal stress associated with water uptake is noticed. Hygrothermal conditioned glass/epoxy composites, when subjected to 70°C temperature and 95% RH to absorb moisture and subjected to 3-point bend tests at 1 and 10 mm/min crosshead speed, show that at a higher strain rate, the mechanical degradation is less pronounced for the same level of absorbed moisture [58].

Ray [59] has shown the effect of temperature on moisture absorption characteristics of carbon fiber/epoxy composites at 60°C temperature and 95% RH and also at 70°C temperature and 95% RH environments. It is clear from the figure that the higher the temperature, the higher the moisture uptake rate (Figure 4.14).

The variations in interlaminar shear strength (ILSS) values for both the types of conditioned specimens are plotted against the percentage of absorbed moisture in Figure 4.15. It is clear that the degree of degradation in shear values is more at a higher conditioning temperature for almost the same amount of absorbed moisture inside the carbon/epoxy laminates for the longer exposure time (i.e., more absorbed moisture). The probable cause for such behavior may be due to the adverse effect of a higher degree of thermal stress at the higher temperature. This higher amount of thermal stress may promote crack initiation and propagation through the boundary layer of high cross-link density at the fiber/matrix interface. The initial rise in ILSS values in both the cases may be due to a strain-free state in the laminates [60,61]. The curing shrinkage stress is released here by the hygroscopic swelling stress in the initial stage of moisture absorption.

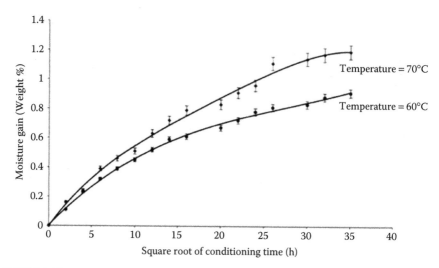

FIGURE 4.14
Moisture uptake kinetics study of carbon/epoxy composites at 60°C and 70°C temperatures, both with 95% RH. (From Ray, B.C., *J. Colloid Interface Sci.*, 298, 111–117, 2006.)

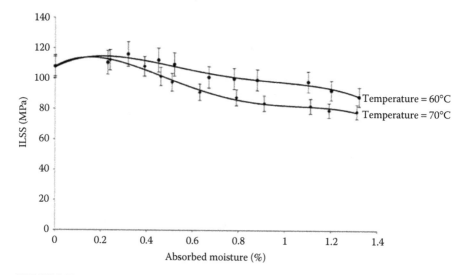

FIGURE 4.15
Effect of conditioning temperature on ILSS of carbon/epoxy composites at various moisture contents. (From Ray, B.C., *J. Colloid Interface Sci.*, 298, 111–117, 2006.)

Figure 4.16 highlights the presence of polymer matrix adherence to the carbon fibers and epoxy matrix damage. The cleaner fibers and interfacial cracking are prevalent in the fractured surface of aged glass/epoxy composites (Figure 4.17). The study reveals that the fiber/matrix adhesive damage and a loss of interfacial integrity are dominating mechanisms in polymer composites during environmental ageing.

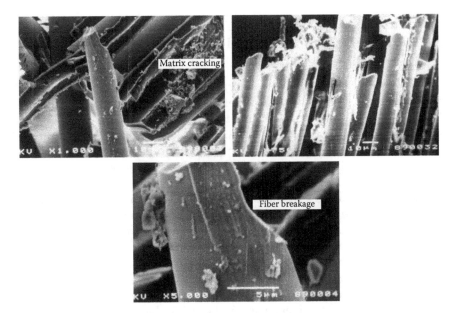

FIGURE 4.16
Scanning electron micrograph shows matrix cracking and fiber damage in carbon/epoxy composites. (From Ray, B.C., *J. Colloid Interface Sci.*, 298, 111–117, 2006.)

FIGURE 4.17
Deadherence and interfacial cracking are evident in aged glass/epoxy laminated composites. (From Ray, B.C., *J. Colloid Interface Sci.*, 298, 111–117, 2006.)

4.4 Fundamentals of Moisture-Induced Degradation Mechanisms

The susceptibility of interphase/interface, which is often termed as "weak link" in the composite, poses a threat on the durability of a FRP composite under a moist environment. In addition, long-term exposure may lead to the degradation of a polymer matrix and fibers that can cause significant damage to the fibrous polymeric composites. An environmental moisture attack can result in various reversible and irreversible physical, chemical, and physico-chemical degradation micromechanisms.

The physical damages in the presence of moisture include swelling and plasticization. The plasticization process, which is schematically shown in Figure 4.18, is the net result of interaction of polymeric chains with water molecules. These interactions have the potential to interrupt the existing hydrogen bonds and can create new hydrogen bonds in the polymeric matrix [63,64]. Alteration in bonding chemistry due to the presence of a water molecule also accounts for the swelling of a composite, which is the net result of increased bond length between polymer chains. Some articles have reported an enhancement in the glass transition temperature (T_g) of polymer during the initial period of water absorption due to the formation of double hydrogen bonds [49,65]. Different physicomechanical phenomena, such as microcrack and microvoid formation also occur in the composite materials, which are not only deleterious for the fibers and matrix, but also for the interface/interphase between them [63,64,66–68]. Formation of these microvoids in the interfacial and matrix region can be attributed to the clustering of water molecules [66]. Moisture-induced internal stresses also termed as "hygrothermal and hydrothermal stresses" may also accelerate the formation of microcracks or microvoids. On the contrary, swelling induced by the moisture may also relieve the residual stresses generated during curing of the matrix. Further, swelling and plasticization are reversible phenomena and material can be recovered after desorption. On the other hand, microcracking and microvoid formation, hydrolysis, leaching, and polymer relaxation are irreversible phenomenon. The micromechanism of hydrolysis includes the detachment of side groups from the backbone of a polymeric

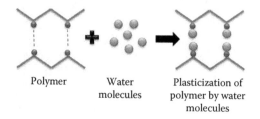

Polymer   Water molecules   Plasticization of polymer by water molecules

FIGURE 4.18
Schematic of Moisture-induced plasticization in polymer. (From Hammiche, D. et al., *Constr. Build. Mater.*, 47, 293–300, 2013.)

chain. Most of the literature have reported that hydrolysis is irreversible, but there are some reports which support that the hydrolyzing effect of diffused water can be reversible [67]. Leaching is another degradation mechanism which may cause the breakdown of the fiber/matrix interface and may also lead to debonding of fibers from the matrix. Structures made up of fibrous polymeric composites are very complex and may fail by a number of micromechanisms, which are not observed in more homogeneous materials [68].

Wang et al. [69] have reported the effect of hygrothermal conditioning (100% RH in a desiccator containing deionized water at the bottom) on the carbon fiber/epoxy composites. The AFM images at various intervals of time have been shown in Figure 4.19. A differential contrast is evolved between

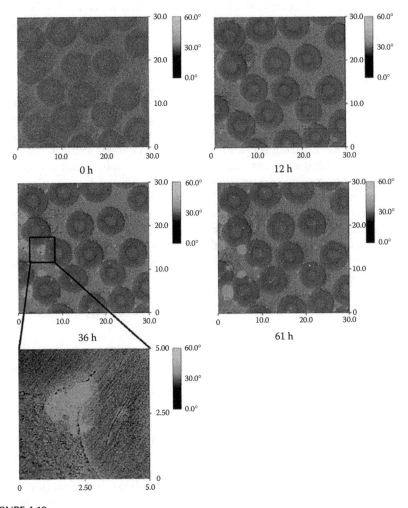

FIGURE 4.19
Hygroscopic treatment effect on the AFM phase images of T700/UF-3369 exposed to 100% RH for different periods of time. (From Wang, Y. and Hahn, T.H., *Compos. Sci. Technol.*, 67, 92–101, 2007.)

the core and the sheath of the carbon fiber after 12 h of humid exposure. The distinction may be thought to be the result of either instrumental artifacts or moisture-induced transformation. Moisture-induced plasticization may help to have a better orientation of the graphitic planes at the sheath region. The evolution of some bright regions at longer exposure times reveals poorly bonded fiber/matrix interface (enlarged view has been shown in Figure 4.19). Hence, the initiation of moisture-induced degradations is likely to be initiated from the interfacial zone.

Fourier transformed infrared spectroscopy is also another important tool for interpreting the impact of absorbed water on the chemistry of the polymer matrix composite [70,71]. Area under the −OH stretching band (at ~3400 cm^{-1}) is very often taken as a measure of water content in the material. Additionally, in the case of epoxy, ring opening polymerization during humid ageing may also be detected by analyzing the epoxide deformation peak at ~916 cm^{-1}. A higher intensity of this peak indicates a less degree of cure in the polymer.

The uncontrolled and nonuniform degradations at micro- and macrolevels manifested in the interphase are because of different environmental conditions during service life. These may restrain its uses in the short-term and also in the long-term reliability of the material. The predicted mechanical behavior may alter during service life because of changes in the nature of the interface. Any changes in the interface might have substantial implications on its performances.

The precise mode of failure is a function of the status of environmentally conditioned interfaces and the time of exposure, thus complicating the prediction of its performances and behavior. The interface is the most highly stressed region of composite materials. The present review highlights the different degrees of damages and degradations under different environmental conditionings. The important roles of interface necessitate a critical and comprehensive understanding of environmentally conditioned interfaces in FRP composite systems.

4.5 Effect of Moisture on Interfacial Durability of Fiber-Reinforced Polymer Composites

Moisture inside a polymer composite leads to matrix plasticization, chemical degradation, and mechanical degradation [7]. Matrix plasticization has deleterious effects on matrix modulus. Chemical degradation includes hydrolysis of bond at the interface. Mechanical degradation is a function of the matrix swelling strain. For Kevlar/epoxy composites, it is observed that 1%

absorbed moisture results in 5%, 4%, and 2% reduction in the compressive, interlaminar shear, and flexural strength, respectively [72]. Similar reductions in the tensile and ILSS of carbon or glass fiber-reinforced polyetherimide have also been reported [73].

Plasticization adversely affects the properties of the polymer composite by inducing plastic deformation in the matrix and by lowering its glass transition temperature. Joshi [60] has investigated the effect of moisture on the ILSS of carbon fiber/epoxy composites. He reported an initial increase in ILSS of about 10% up to 0.1 weight% absorbed moisture and a subsequent decrease by 25% at maximum moisture of approximately 2%.

The effect of water sorption on mechanical behavior FRP composites has been investigated by many researchers [74–76]. It was reported that in the case of woven glass/epoxy composites delamination, load-carrying capability was reduced to 40% with 1.29% absorbed moisture [74].

Akay [75] reported that static and fatigue strength of a carbon fiber/epoxy composite decreases when subjected to hygrothermal conditioning. The reduction in mechanical properties after hygrothermal ageing was reported to be dependent on the ply stacking sequence in the laminates. The maximum reduction in static bearing strength (22%) was observed for a block ply arrangement, that is $(\pm 45°)_2/(0°,90°)_4/(\pm 45°)_2$, whereas a slight increment was recorded for the $(0°,90°)_{14}$ laminate. This increment was attributed to the moisture-induced plasticization, which may have reduced the extent of local stress concentrations induced by 0°/90° ply arrangements. Lassila et al. [76] have reported a reduction in flexural strength of E-glass FRP composites when exposed to water for 30 days.

E-glass/vinyl ester composites submerged in freshwater for a period of about 2 years, showed 60% and 10% reductions in tensile strength and stiffness, respectively [77]. ILSS reduction due to moisture absorption was also observed for aramid/epoxy composites [78]. Ageing of graphite-fabric epoxy composites in boiling water even after a relatively short exposure of 46 h causes a large reduction in both shear and tensile strengths [79]. For symmetric and anti-symmetric GFRP laminates, reductions in flexural stiffness were reported as 54% and 27%, respectively, when exposed to a 98% humid environment for 2000 h [80]. Depending on the lay-up sequence, the maximum static bearing strength is reduced by 8% to 22% for woven laminates. Glass fiber (with and without solvent) and carbon fiber- (with and without solvent) reinforced composite laminates were characterized in terms of water sorption and desorption [50]. The total amount of water absorbed for a 1,200 h exposure at 71°C was 4%–5% of the total laminate weight and after 450 h desorption water content was approximately 3%. Poor interfacial adhesion and change in failure mode due to moisture accumulation in the composite are regarded as the main cause of large reductions in flexural and interlaminar strength values [81,82].

References

1. A. Boisseau, P. Davies, F. Thiebaud, Sea water ageing of composites for ocean energy conversion systems: Influence of glass fibre type on static behaviour, *Appl. Compos. Mater.*, 19: 459–473, 2012. doi:10.1007/s10443-011-9219-6.
2. J. Hoshikuma, K. Kawashima, K. Nagaya, A. W. Taylor, Stress-strain model for confined reinforced concrete in bridge piers, *J. Struct. Eng.*, 123: 624–633, 1997.
3. J. D. Garcia-Espinel, D. Castro-Fresno, P. Parbole Gayo, F. Ballester-Muñoz, Effects of sea water environment on glass fiber reinforced plastic materials used for marine civil engineering constructions, *Mater. Des.*, 66(Part A): 46–50, 2015. doi:10.1016/j.matdes.2014.10.032.
4. G. Marom, L. J. Broutman. Moisture penetration into composites under external stress, *Polym. Compos.*, 2: 132–136, 1981. doi:10.1002/pc.750020310.
5. D. H. Kaelble, P. J. Dynes, L. Maus, Hydrothermal aging of composite materials part 1: Interfacial aspects, *J. Adhes.*, 8: 121–144, 1976. doi:10.1080/00218467608075078.
6. B. C. Ray, Effects of crosshead velocity and sub-zero temperature on mechanical behaviour of hygrothermally conditioned glass fibre reinforced epoxy composites, *Mater. Sci. Eng. A*, 379: 39–44, 2004. doi:10.1016/j.msea.2003.11.031.
7. C.-H. Shen, G. S. Springer, Moisture absorption and desorption of composite materials, *J. Compos. Mater.*, 10: 2–20, 1976. doi:10.1177/002199837601000101.
8. A. Chateauminois, L. Vincent, B. Chabert, J. P. Soulier, Study of the interfacial degradation of a glass-epoxy composite during hygrothermal ageing using water diffusion measurements and dynamic mechanical thermal analysis, *Polymer*, 35: 4766–4774, 1994. doi:10.1016/0032-3861(94)90730-7.
9. H. S. Choi, K. J. Ahn, J.-D. Nam, H. J. Chun, Hygroscopic aspects of epoxy/carbon fiber composite laminates in aircraft environments, *Compos. Part A Appl. Sci. Manuf.*, 32: 709–720, 2001. doi:10.1016/S1359-835X(00)00145-7.
10. F. Ellyin, R. Maser, Environmental effects on the mechanical properties of glass-fiber epoxy composite tubular specimens, *Compos. Sci. Technol.*, 64: 1863–1874, 2004. doi:10.1016/j.compscitech.2004.01.017.
11. M. Assarar, D. Scida, A. El Mahi, C. Poilâne, R. Ayad, Influence of water ageing on mechanical properties and damage events of two reinforced composite materials: Flax-fibres and glass-fibres, *Mater. Des.*, 32: 788–795, 2011. doi:10.1016/j.matdes.2010.07.024.
12. D. Scida, M. Assarar, C. Poilâne, R. Ayad, Influence of hygrothermal ageing on the damage mechanisms of flax-fibre reinforced epoxy composite, *Compos. Part B Eng.*, 48: 51–58, 2013. doi:10.1016/j.compositesb.2012.12.010.
13. A. Zafar, F. Bertocco, J. Schjødt-Thomsen, J. C. Rauhe, Investigation of the long term effects of moisture on carbon fibre and epoxy matrix composites, *Compos. Sci. Technol.*, 72: 656–666, 2012. doi:10.1016/j.compscitech.2012.01.010.
14. G. Mensitieri, M. Lavorgna, P. Musto, G. Ragosta, Water transport in densely crosslinked networks: A comparison between epoxy systems having different interactive characters, *Polymer*, 47: 8326–8336, 2006. doi:10.1016/j.polymer.2006.09.066.
15. M. Al-Harthi, K. Loughlin, R. Kahraman, Moisture diffusion into epoxy adhesive: Testing and modeling, *Adsorption*, 13: 115–120, 2007. doi:10.1007/s10450-007-9011-y.

16. T. Glaskova, A. Aniskevich, Moisture absorption by epoxy/montmorillonite nanocomposite, *Compos. Sci. Technol.*, 69: 2711–2715, 2009. doi:10.1016/j.compscitech.2009.08.013.

17. L.-R. Bao, A. F. Yee, Moisture diffusion and hygrothermal aging in bisma-leimide matrix carbon fiber composites—part I: Uni-weave composites, *Compos. Sci. Technol.*, 62: 2099–2110, 2002. doi:10.1016/S0266-3538(02)00161-6.

18. S. Cotugno, G. Mensitieri, P. Musto, L. Sanguigno, Molecular interactions in and transport properties of densely cross-linked networks: A time-resolved FT-IR spectroscopy investigation of the epoxy/H_2O system, *Macromolecules*, 38: 801–811, 2005. doi:10.1021/ma040008j.

19. W. Liu, S. V. Hoa, M. Pugh, Water uptake of epoxy-clay nanocomposites: Model development, *Compos. Sci. Technol.*, 67: 3308–3315, 2007. doi:10.1016/j.compscitech.2007.03.041.

20. G. LaPlante, A. V. Ouriadov, P. Lee-Sullivan, B. J. Balcom, Anomalous mois-ture diffusion in an epoxy adhesive detected by magnetic resonance imaging, *J. Appl. Polym. Sci.*, 109: 1350–1359, 2008. doi:10.1002/app.28106.

21. L. R. Grace, M. C. Altan, Characterization of anisotropic moisture absorption in polymeric composites using hindered diffusion model, *Compos. Part A: Appl. Sci. Manuf.*, 43: 1187–1196, 2012. doi:10.1016/j.compositesa.2012.03.016.

22. H. G. Carter, K. G. Kibler, Langmuir-type model for anomalous moisture diffu-sion in composite resins, *J. Compos. Mater.*, 12: 118–131, 1978.

23. V. La Saponara, Environmental and chemical degradation of carbon/epoxy and structural adhesive for aerospace applications: Fickian and anomalous dif-fusion, Arrhenius kinetics, *Compos. Struct.*, 93: 2180–2195, 2011. doi:10.1016/j.compstruct.2011.03.005.

24. L. Kumosa, B. Benedikt, D. Armentrout, M. Kumosa, Moisture absorption properties of unidirectional glass/polymer composites used in composite (non-ceramic) insulators, *Compos. Part A Appl. Sci. Manuf.*, 35: 1049–1063, 2004. doi:10.1016/j.compositesa.2004.03.008.

25. L. Kumosa, D. Armentrout, B. Benedikt, M. Kumosa, An investigation of mois-ture and leakage currents in GRP composite hollow cylinders, *IEEE Trans. Dielectr. Electr. Insul.*, 12: 1043–1059, 2005. doi:10.1109/TDEI.2005.1522197.

26. D. Perreux, C. Suri, A study of the coupling between the phenomena of water absorption and damage in glass/epoxy composite pipes, *Compos. Sci. Technol.*, 57: 1403–1413, 1997. doi:10.1016/S0266-3538(97)00076-6.

27. I. Merdas, F. Thominette, A. Tcharkhtchi, J. Verdu, Factors governing water absorption by composite matrices, *Compos. Sci. Technol.*, 62: 487–492, 2002. doi:10.1016/S0266-3538(01)00138-5.

28. G. Kotsikos, A. G. Gibson, J. Mawella, Assessment of moisture absorption in marine GRP laminates with aid of nuclear magnetic resonance imaging, *Plast Rubber Compos.*, 36: 413–418, 2007. doi:10.1179/174328907×248203.

29. L.-R. Bao, A. F. Yee, Effect of temperature on moisture absorption in a bis-maleimide resin and its carbon fiber composites, *Polymer*, 43: 3987–3997, 2002. doi:10.1016/S0032-3861(02)00189-1.

30. P. M. Jacobs, F. R. Jones, Diffusion of moisture into two-phase polymers, *J. Mater. Sci.*, 24: 2331–2336, 1989. doi:10.1007/BF01174492.

31. X. Tang, J. D. Whitcomb, Y. Li, H.-J. Sue, Micromechanics modeling of mois-ture diffusion in woven composites, *Compos. Sci. Technol.*, 65: 817–826, 2005. doi:10.1016/j.compscitech.2004.01.015.

32. M. D. Placette, X. Fan, J.-H. Zhao, D. Edwards, Dual stage modeling of moisture absorption and desorption in epoxy mold compounds, *Microelectron. Reliab.*, 52: 1401–1408, 2012. doi:10.1016/j.microrel.2012.03.008.

33. A. R. Berens, H. B. Hopfenberg, Diffusion and relaxation in glassy polymer powders: 2. Separation of diffusion and relaxation parameters, *Polymer*, 19: 489–496, 1978. doi:10.1016/0032-3861(78)90269-0.

34. X. Jiang, H. Kolstein, F. S. K. Bijlaard, Moisture diffusion in glass–fiber-reinforced polymer composite bridge under hot/wet environment, *Compos. Part B Eng.*, 45: 407–416, 2013. doi:10.1016/j.compositesb.2012.04.067.

35. V. M. Karbhari, G. Xian, Hygrothermal effects on high VF pultruded unidirectional carbon/epoxy composites: Moisture uptake, *Compos. Part B Eng.*, 40: 41–49, 2009. doi:10.1016/j.compositesb.2008.07.003.

36. A. Apicella, L. Nicolais, Effect of water on the properties of epoxy matrix and composite, *Epoxy Resins Compos. I*, Springer, pp. 69–77, 1985.

37. C. Carfagna, A. Apicella, Physical degradation by water clustering in epoxy resins, *J. Appl. Polym. Sci.*, 28: 2881–2885, 1983. doi:10.1002/app.1983.070280917.

38. A. C. Karmaker, Effect of water absorption on dimensional stability and impact energy of jute fibre reinforced polypropylene, *J. Mater. Sci. Lett.*, 16: 462–464, 1997. doi:10.1023/A:1018508209022.

39. M. M. Thwe, K. Liao, Effects of environmental aging on the mechanical properties of bamboo–glass fiber reinforced polymer matrix hybrid composites, *Compos. Part A Appl. Sci. Manuf.*, 33: 43–52, 2002. doi:10.1016/S1359-835X(01)00071-9.

40. E. N. Brown, A. K. Davis, K. D. Jonnalagadda, N. R. Sottos, Effect of surface treatment on the hydrolytic stability of E-glass fiber bundle tensile strength, *Compos. Sci. Technol.*, 65: 129–136, 2005. doi:10.1016/j.compscitech.2004.07.001.

41. C. L. Schutte, Environmental durability of glass-fiber composites, *Mater. Sci. Eng. R Rep.*, 13: 265–323, 1994.

42. O. Starkova, S. T. Buschhorn, E. Mannov, K. Schulte, A. Aniskevich, Water transport in epoxy/MWCNT composites, *Eur. Polym. J.*, 49: 2138–2148, 2013. doi:10.1016/j.eurpolymj.2013.05.010.

43. C. L. Soles, A. F. Yee, A discussion of the molecular mechanisms of moisture transport in epoxy resins, *J. Polym. Sci. Part B Polym. Phys.*, 38: 792–802, 2000.

44. X. Fan, E. Suhir, *Moisture Sensitivity of Plastic Packages of IC Devices*. New York: Springer, 2010.

45. A. Apicella, L. Nicolais, C. de Cataldis, Characterization of the morphological fine structure of commercial thermosetting resins through hygrothermal experiments, in *Charact. Polym. Solid State I Part A NMR Spectrosc. Methods Part B Mech. Methods*, H. H. Kaush, H. G. Zachman (Eds.). Berlin, Germany: Springer, pp. 189–207, 1985. doi:10.1007/3-540-13779-3_21.

46. P. Nogueira, C. Ramírez, A. Torres, M. J. Abad, J. Cano, J. López et al., Effect of water sorption on the structure and mechanical properties of an epoxy resin system, *J. Appl. Polym. Sci.*, 80: 71–80, 2001.

47. E. Gaudichet-Maurin, F. Thominette, J. Verdu, Water sorption characteristics in moderately hydrophilic polymers, Part 1: Effect of polar groups concentration and temperature in water sorption in aromatic polysulfones, *J. Appl. Polym. Sci.*, 109: 3279–3285, 2008. doi:10.1002/app.24873.

48. S. Popineau, C. Rondeau-Mouro, C. Sulpice-Gaillet, M. E. R. Shanahan, Free/bound water absorption in an epoxy adhesive, *Polymer*, 46: 10733–10740, 2005. doi:10.1016/j.polymer.2005.09.008.

49. J. Zhou, J. P. Lucas, Hygrothermal effects of epoxy resin. Part I: the nature of water in epoxy, *Polymer*, 40: 5505–5512, 1999. doi:10.1016/S0032-3861(98)00790-3.

50. F. U. Buehler, J. C. Seferis, Effect of reinforcement and solvent content on moisture absorption in epoxy composite materials, *Compos. Part A Appl. Sci. Manuf.*, 31: 741–748, 2000. doi:10.1016/S1359-835X(00)00036-1.

51. P. Moy, F. E. Karasz, Epoxy-water interactions, *Polym. Eng. Sci.*, 20: 315–319, 1980. doi:10.1002/pen.760200417.

52. G. J. Fleming, T. Rose, Graphite composites for ocean ordinances, US Department of Commerce, National Technical Information Service Report AD-770-407, 1973.

53. J. Scheirs, *Compositional and Failure Analysis of Polymers: A Practical Approach.* Chichester, UK: John Wiley & Sons, 2000.

54. D. Olmos, R. López-Morón, J. González-Benito, The nature of the glass fibre surface and its effect in the water absorption of glass fibre/epoxy composites, The use of fluorescence to obtain information at the interface, *Compos. Sci. Technol.*, 66: 2758–2768, 2006. doi:10.1016/j.compscitech.2006.03.004.

55. J. G. Iglesias, J. González-Benito, A. J. Aznar, J. Bravo, J. Baselga, Effect of glass fiber surface treatments on mechanical strength of epoxy based composite materials, *J. Colloid Interface Sci.*, 250: 251–260, 2002.

56. B. C. Ray, D. Rathore, Environmental damage and degradation of FRP composites: A review report, *Polym. Compos.*, 36(3): 410–423, 2014. doi:10.1002/pc.22967.

57. B. C. Ray, S. Mula, T. Bera, P. K. Ray, Prior thermal spikes and thermal shocks on mechanical behavior of glass fiber-epoxy composites, *J. Reinf. Plast Compos.*, 25: 197–213, 2006. doi:10.1177/0731684405056446.

58. B. C. Ray, Effect of hydrothermal shock cycles on shear strength of glass fiber-polyester composites, *J. Reinf. Plast Compos.*, 24: 1335–1340, 2005. doi:10.1177/0731684405049854.

59. B. C. Ray, Temperature effect during humid ageing on interfaces of glass and carbon fibers reinforced epoxy composites, *J. Colloid Interface Sci.*, 298: 111–117, 2006. doi:10.1016/j.jcis.2005.12.023.

60. O. K. Joshi, The effect of moisture on the shear properties of carbon fibre composites, *Composites*, 14: 196–200, 1983. doi:10.1016/0010-4361(83)90005-8.

61. B. C. Ray, A. Biswas, P. K. Sinha, Freezing and thermal spikes effects on interlaminar shear strength values of hygrothermally conditioned glass fibre/epoxy composites, *J. Mater. Sci. Lett.*, 11: 508–509, 1992. doi:10.1007/BF00731120.

62. D. Hammiche, A. Boukerrou, H. Djidjelli, Y.-M. Corre, Y. Grohens, I. Pillin, Hydrothermal ageing of alfa fiber reinforced polyvinylchloride composites, *Constr. Build. Mater.*, 47: 293–300, 2013. doi:10.1016/j.conbuildmat.2013.05.078.

63. A. H. Nissan, H-bond dissociation in hydrogen bond dominated solids, *Macromolecules*, 9: 840–850, 1976. doi:10.1021/ma60053a026.

64. J. Mijović, H. Zhang, Molecular dynamics simulation study of motions and interactions of water in a polymer network, *J. Phys. Chem. B*, 108: 2557–2563, 2004. doi:10.1021/jp036181j.

65. L. Li, S. Zhang, Y. Chen, M. Liu, Y. Ding, X. Luo et al., Water transportation in epoxy resin, *Chem. Mater.*, 17: 839–845, 2005. doi:10.1021/cm048884z.

66. H. Ishida, J. L. Koenig, The reinforcement mechanism of fiber-glass reinforced plastics under wet conditions: A review, *Polym. Eng. Sci.*, 18: 128–145, 1978. doi:10.1002/pen.760180211.

67. M. K. Antoon, J. L. Koenig, Structure and moisture stability of the matrix phase in glass-reinforced epoxy composites, *J. Macromol. Sci. Rev. Macromol. Chem.*, C19: 135–173, 1980.

68. B. C. Ray, D. Rathore, Durability and integrity studies of environmentally conditioned interfaces in fibrous polymeric composites: Critical concepts and comments, *Adv. Colloid Interface Sci.*, 209: 68–83, 2014. doi:10.1016/j.cis.2013.12.014.

69. Y. Wang, T. H. Hahn, AFM characterization of the interfacial properties of carbon fiber reinforced polymer composites subjected to hygrothermal treatments, *Compos. Sci. Technol.*, 67: 92–101, 2007. doi:10.1016/j.compscitech.2006.03.030.

70. C. D. Arvanitopoulos, J. L. Koenig, Infrared spectral imaging of the interphase of epoxy-glass fiber-reinforced composites under wet conditions, *Appl. Spectrosc.*, 50: 11–18, 1996.

71. W. Noobut, J. L. Koenig, Interfacial behavior of epoxy/E-glass fiber composites under wet-dry cycles by fourier transform infrared microspectroscopy, *Polym. Compos.*, 20: 38–47, 1999. doi:10.1002/pc.10333.

72. M. Akay, S. K. Ah Mun, A. Stanley, Influence of moisture on the thermal and mechanical properties of autoclaved and oven-cured Kevlar-49/epoxy laminates, *Compos. Sci. Technol.*, 57: 565–571, 1997. doi:10.1016/S0266-3538(97)00017-1.

73. J. Viña, E. A. García, A. Argüelles, I. Viña, The effect of moisture on the tensile and interlaminar shear strengths of glass or carbon fiber reinforced PEI, *J. Mater. Sci. Lett.*, 19: 579–581, 2000. doi:10.1023/A:1006778228209.

74. T. A. Collings, D. E. W. Stone, Hygrothermal effects in CFC laminates: Damaging effects of temperature, moisture and thermal spiking, *Compos. Struct.*, 3: 341–378, 1985. doi:10.1016/0263-8223(85)90061-3.

75. M. Akay, Bearing strength of as-cured and hygrothermally conditioned carbon fibre/epoxy composites under static and dynamic loading, *Composites*, 23: 101–108, 1992. doi:10.1016/0010-4361(92)90110-G.

76. L. V. J. Lassila, T. Nohrström, P. K. Vallittu, The influence of short-term water storage on the flexural properties of unidirectional glass fiber-reinforced composites, *Biomaterials*, 23: 2221–2229, 2002. doi:10.1016/S0142-9612(01)00355-6.

77. S. P. Phifer, Quasi-static and fatigue evaluation of pultruded vinyl ester/E-glass composites. Virginia Tech, 1998.

78. L. E. Doxsee, W. Janssens, I. Verpoest, P. De Meester, Strength of aramid-epoxy composites during moisture absorption, *J. Reinf. Plast Compos.*, 10: 645–655, 1991. doi:10.1177/073168449101000606.

79. S. Birger, A. Moshonov, S. Kenig, The effects of thermal and hygrothermal ageing on the failure mechanisms of graphite-fabric epoxy composites subjected to flexural loading, *Composites*, 20: 341–348, 1989. doi:10.1016/0010-4361(89)90659-9.

80. P. K. Aditya, P. K. Sinha, Diffusion coefficients of polymeric composites subjected to periodic hygrothermal exposure, *J. Reinf. Plast Compos.*, 11: 1035–1047, 1992. doi:10.1177/073168449201100904.

81. J. M. Whitney, G. E. Husman, Use of the flexure test for determining environmental behavior of fibrous composites, *Exp. Mech.*, 18: 185–190, 1978. doi:10.1007/BF02324140.

82. L. T. Drzal, M. Madhukar, Fibre-matrix adhesion and its relationship to composite mechanical properties, *J. Mater. Sci.*, 28: 569–610, 1993. doi:10.1007/BF01151234.

5

Hygrothermal-Dominated Failure in Polymer Matrix Composites

5.1 Introduction

The kinetics of water/moisture ingression and its subsequent effects on the structural durability of PMC have been discussed in Chapter 4. In this chapter, focus has been given on the effects of some of the dynamic conditioning and testing parameters on the mechanical performance of hygrothermally conditioned fiber-reinforced polymer (FRP) composites.

5.2 Freezing of Absorbed Moisture

When a FRP composite is exposed to a hygrothermal environment, it may lead to polymer degradation, interfacial debonding, and eventually gross delamination of the material. In various practical applications, environmental moisture is absorbed by the FRP composite during its service. This ends up with reduced stress transfer efficiency from the polymer to the fiber, polymer plasticization, and so on. The induced swelling strain influences this degradation to a significant extent, as this tries to detach the matrix from the fiber [1]. Further, in-service application may exhibit a sub-zero temperature at which the absorbed moisture tends to freeze. Ray [2] has investigated the effect of freezing of absorbed moisture on the interfacial durability of glass FRP composites. The glass/epoxy composite was initially conditioned at 60°C and 95% relative humidity (RH) for different time periods. Subsequently, the conditioned samples were put in a deep freezer maintained at −6°C for 24 h. A short beam shear test was then conducted on moisture conditioned samples with and without freezing to find the effect of frozen moisture on the interlaminar shear strength (ILSS) of the material.

Effect of conditioning time on the ILSS of the glass/epoxy composite has been plotted in Figure 5.1. After each of this conditioning time, the absorbed moisture was frozen and then tested. The results thus obtained have also been plotted in Figure 5.1.

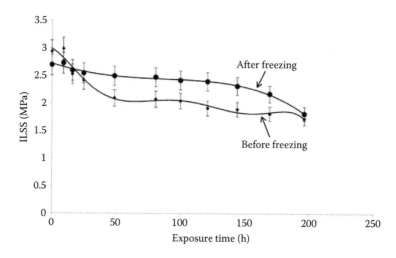

FIGURE 5.1
Effect of freezing of the absorbed moisture on the ILSS of glass/epoxy composites. (From Ray, B.C., *Mater. Sci. Eng. A*, 379, 39–44, 2004.)

Hygroscopic stress is generated at the fiber/polymer interface due to differential moisture absorption causing differential swelling. This, along with polymer plasticization, reduces the ILSS of the composite, as the conditioning time/moisture content was increased. After this hygrothermal treatment, when the absorbed moisture was frozen, some extra residual stresses have been generated. This is due to the volumetric expansion of the absorbed moisture as it solidifies. Consequently, the ILSS of the composite is altered as can be seen from Figure 5.1.

5.3 Effect of Loading Rate

Ray [2] has also studied the effect of the loading rate on the ILSS of a hygrothermally conditioned glass/epoxy composite. After exposing the samples at 60°C and 95% RH for different time lengths, a short beam shear test was conducted at two different loading rates, 1 and 10 mm/min. The results thus obtained are presented in Figure 5.2.

Irrespective of the loading rate, there is a reduction in ILSS of the composite with an increase in conditioning time. However, with the same conditioning time, the ILSS was found to be slightly lower for a 1 mm/min than a 10 mm/min loading rate. The higher time available for the absorbed water to interact with the matrix/interface during testing causes more deterioration in the ILSS of the samples.

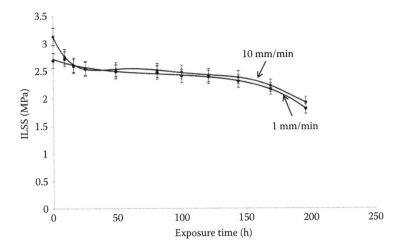

FIGURE 5.2
Effect of loading rate on the ILSS of hygrothermally conditioned glass/epoxy composites.
(From Ray, B.C., *Mater. Sci. Eng. A*, 379, 39–44, 2004.)

5.4 Effect of Hygrothermal Cycling

In practice, there exists a number of applications where there is a fluctuation in the temperature and/or RH of the in-service environment. Hence, efforts have been made to assess the effect of these changing environments on the performance of the FRP composites.

5.4.1 Thermal Fatigue

Thermal fatigue during hygrothermal exposure refers to a state where the conditioning temperature fluctuates, but the RH remains constant. Ray [3] had designed a hygrothermal cycle consisting of 50°C and 60% RH for 1 h followed by 70°C and 60% RH for next 1 h. This cycle was repeated 19 times on a glass/polyester composite, and the results obtained are plotted in Figure 5.3.

Repeated changes in the moisture absorption profile of the material significantly alter the interfacial durability of FRP composites. The diffusion of water inside and outside the polymer matrix and its accumulation at the fiber/polymer interface is responsible for the changes in mechanical performance of these composites. Nonuniform distribution of moisture in the bulk material and moisture-induced physical, chemical, and physicochemical changes in the material brings changes in the mechanical properties of FRP composites.

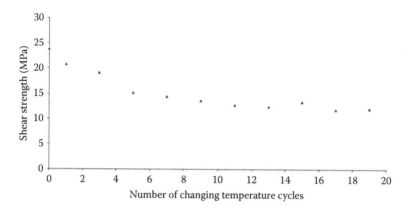

FIGURE 5.3
Effect of hygrothermal temperature cycles on ILSS of glass/polyester composites. (From Ray, B.C., *J. Reinf. Plast. Compos.*, 25, 1227–1240, 2006.)

5.4.2 Relative Humidity Cycling

Similar to that of thermal fatigue during hygrothermal conditioning, if the RH during hygrothermal conditioning is changed repeatedly keeping the temperature constant, it is called RH cycling during hygrothermal conditioning. Ray [3] has done hygrothermal cycling of glass/epoxy composites by conditioning the samples at 50°C and 60% RH for 1 h and the next hour at 50°C and 95% RH. This cycle was repeated 19 times, and the results thus obtained are reported in Figure 5.4.

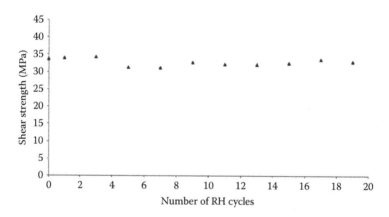

FIGURE 5.4
Effect of hygrothermal RH cycles on ILSS of glass/epoxy composites. (From Ray, B.C., *J. Reinf. Plast. Compos.*, 25, 1227–1240, 2006.)

5.5 Summary

FRP composite has an inherent tendency of absorbing environmental moisture. Further, most of the experimental hygrothermal ageing of composites has been done at a constant temperature or RH. However, some research has also been done where there is a cyclic variation in the temperature and/ or RH during conditioning, which is very practical in various applications. Further, there may be a number of subsequent events that may take place after moisture absorption, such as the freezing of absorbed moisture, which may influence the behavior of the FRP composites. Further research of this kind should be conducted to simulate the practical conditions more precisely, so as to predict the long-term behavior of the material.

References

1. M. S. Amer, M. J. Koczak, and L. S. Schadler, Environmental degradation of the interface in graphite/epoxy single filament composites measured using Raman spectroscopy: Effect of hydrothermal and thermal exposure at 100°C, *Compos. Interfaces*, 3: 41–50, 1995. doi:10.1163/156855495x00147.
2. B. C. Ray, Effects of crosshead velocity and sub-zero temperature on mechanical behaviour of hygrothermally conditioned glass fibre reinforced epoxy composites, *Mater. Sci. Eng. A*, 379: 39–44, 2004. doi:10.1016/j.msea.2003.11.031.
3. B. C. Ray, Effects of changing environment and loading speed on mechanical behavior of FRP composites. *J. Reinf. Plast. Compos.*, 25: 1227–1240, 2006. doi:10.1177/0731684406059783.

6

Low Earth Orbit Space Environmental- and Other Environmental-Dominated Failure in Polymer Matrix Composites

Fiber reinforced polymer (FRP) composites are prime choice materials for various space structures and systems due to their outstanding thermal, optical, mechanical and electrical properties. High specific strength and stiffness coupled with excellent fatigue resistance and low coefficient of thermal expansion (CTE) enable them as candidate materials which fulfill the design requirements of space shuttle, spacecrafts, launch vehicles and space stations. Utilization of FRP to design and manufacture of space structural components has also an advantage of relatively less structure assembly time and overall cost. Despite of numerous advantages over metallic counterparts, if the structural requirement is to operate for prolonged duration in the space environment, FRP composites may have difficulty in maintaining their outstanding performance against the harsh space environment especially in low earth orbit (LEO) space environment. The space environment includes UV and atomic oxygen (AO) irradiation, very high vacuum and space debris with very high velocity. The components of space environment are very hazardous to FRP composites. This chapter provides an insight on the effects of different components of space environment on the durability of FRP composites.

6.1 Effect of Ultraviolet Radiations on Fiber-Reinforced Polymers

Ultraviolet (UV) has a wavelength between 290 and 400 nm, and the energy associated with these wavelengths is equivalent to the bond energy of the polymeric materials. Hence, these wavelengths can dissociate the molecule bonds in polymers and may lead to the degradation of the materials [1]. The degradation starts at the outer surface of the polymeric materials, which is exposed to the UV light. If the light penetration is limited to the surface only, then it may result in surface discoloration only, and if the degradation penetrates through the bulk of the material, then it can result in degradation in the mechanical properties of the corresponding polymer materials [1].

Exposure to UV radiation causes hydrogen abstraction from the polymer molecules, which generates free radicals, and these free radicals can initiate other reactions causing brittleness and, subsequently, lower the molecular weights and loss of thermal diffusivity and load-bearing capacity [2–4]. The extent of degradation of the material under the exposure of UV radiation depends on the type of polymer and the duration of exposure [2,5]. UV radiation can cause either random chain scissions or increase in cross-linking density in the polymers and consequently leads to variations in brittleness and reduction in material strength [4]. Phelps and Long [5] reported no decrease in flexural properties of irradiated composites under short-term exposure. Short-term UV exposure resulted in changes in the surface morphology of the composites. Fritz Larsson [6] studied the effect of UV light on the mechanical properties of Kevlar 49 composites. Their results elucidated that the degradation depends on the thickness of the composite. They postulated that only 0.13 mm thick specimens were found affected by UV exposure, and their strength retained 60% after 1,000 h of exposure. No degradation effect on 0.25 and 0.50 mm thick laminates was observed. It is also observed that the damage due to UV radiation is higher when the composites are exposed to air than in a near vacuum system [7]. The elastic modulus of a single Kevlar fiber exposed to UV irradiation and water increased 30% and decreased 15% compared to that of the untreated ones, respectively, as shown in Figure 6.1 [8].

Scaffaro et al. [9] demonstrated new equipment to measure the combined effects of humidity, temperature, mechanical stress, and UV exposure on

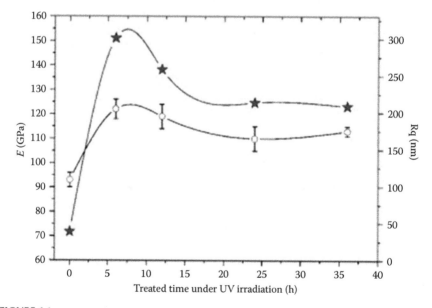

FIGURE 6.1
Effect of UV irradiation time on the elastic modulus and surface roughness of Kevlar fiber. (From Wang, H. et al., *Polym. Degrad. Stab.*, 97, 1755–1761, 2012.)

the creep behavior of polymers. Surface molecular characterization of epoxy resin composites reinforced with E-glass, three dimensional (3D) glass, and carbon fiber subjected to an intense UV and high temperature environment has been carried out using X-ray photoelectron spectroscopy (XPS) and Time-of-Flight-Secondary Ion Mass Spectrometry (ToF–SIMS) [10]. The XPS results revealed that 3D-glass reinforced composites exhibit more surface oxidation in comparison with E-glass and carbon fiber composites. Principal components analysis of the ToF–SIMS positive ion spectra indicated that E-glass and 3D-glass reinforced composites suffered chain scission, while carbon fiber composites suffered chain scission and cross-linking reactions under UV excursion (Figure 6.2).

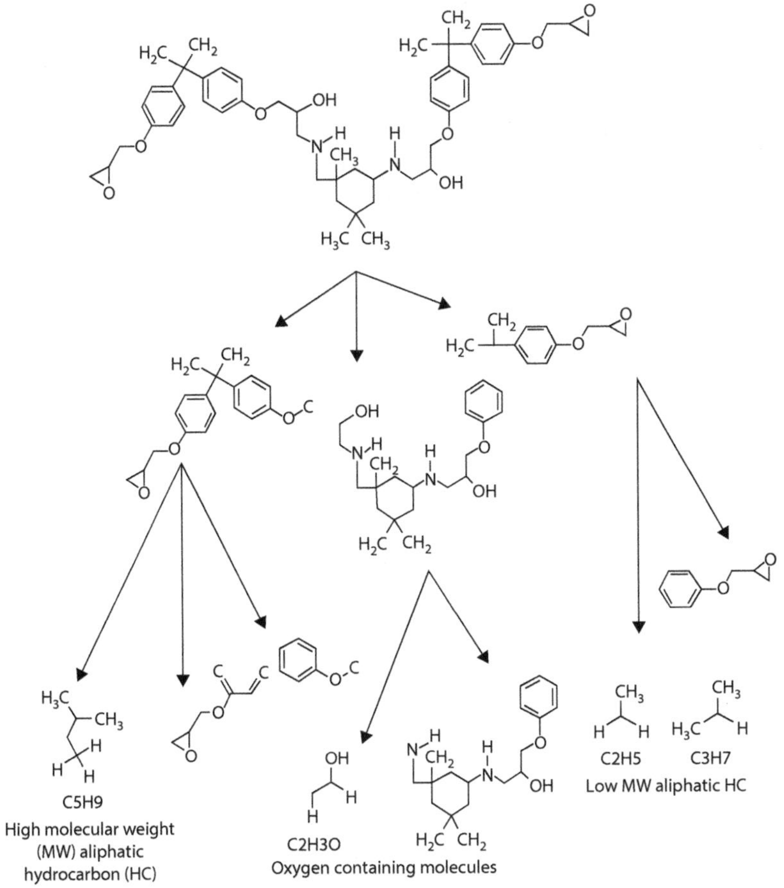

FIGURE 6.2
Schematic diagram of proposed chain scission reaction of diglycidyl ether of bisphenol A/ isophorone diamine (DGEBA/IPD) cured resin matrix as per the results extracted from ToF–SIMS data. (From Awaja, F. and Pigram, P. J., *Polym. Degrad. Stab.*, 94, 651–658, 2009.)

6.2 Effects of Vacuum Thermal Cycling

An investigation on the effect of vacuum thermal cycling on the physical properties of a unidirectional M40J/AG-80 composite revealed that by increasing the thermal cycles, both the 90° and 0° tensile strengths decrease [11]. Initially, an increase in bend strength is noticed, but after 97 cycles, it was decreased to a certain extent. The mass loss ratio during vacuum thermal cycling increases to a given value after 48 cycles, and this mass loss is attributed to the volatilization of smaller molecules from the composite. Park et al. [12] also studied the effect of vacuum thermal cycling on the properties of unidirectional carbon fiber/epoxy composites. Vacuum thermal cycling was performed under the high-vacuum state of 1.3^{-3} Pa in the temperature range between 120°C and 175°C for up to 2,000 cycles. The degradation in the properties of composites under vacuum thermal cycling is attributed to the formation of microvoids and interfacial sliding at the fiber/matrix interface [12]. Interface dominated properties, such as compressive strength and interlaminar shear strength (ILSS), were remarkably reduced up to 15% [12]. The variation in mechanical properties has been shown in Figure 6.3. Baluch et al. [13] studied the effect of hypervelocity impact on CFRP composites under LEO space environment. The objective was to evaluate the potential of CFRP composites as a candidate material for application in bumper of the spacecraft. They suggested that the the space shielding concept can be effectively utilized by making the shielding system made of composites and inclined in orientation with respect to the space debris attacks.

6.3 Irradiation Induced Damages

The exposure of composite materials to extreme radiation environments, however, may result in damage of the structure and microstructure. Especially, neutron irradiation has been noticed to cause delamination in stratified structures or debonding at fiber/polymer interfaces in organic composites. Irradiation can also change the electrical and optical properties of the graphite [14]. Excursion of carbon materials to ionizing radiation can modify the crystal lattice by displacement of atoms within the lattice or electronic excitation [15]. Lattice defects in carbon materials can produce a variety of effects, and these defects can lead to the modifications in most physical properties, including thermal and mechanical properties. Generally, it is observed that the mechanical properties of carbon materials have not been affected to any great extent by the electronic excitation. However, electrons ejected from atoms create the active sites that are sensitive to environment

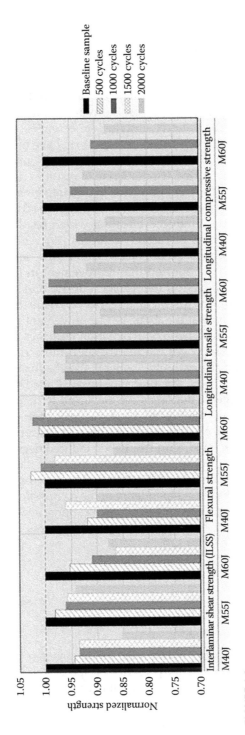

FIGURE 6.3
Mechanical property variations of the three kinds of unidirectional carbon fiber/epoxy laminate as a function of vacuum thermal cycling. (From Park, S. Y. et al., *Compos. Part B Eng.*, 43, 726–738, 2012.)

species and can accelerate other damage mechanisms, namely moisture uptake [15]. Irradiation with fast neutrons changes the value of interlaminar shear strength as observed in carbon/carbon composite [16]. Other studies have shown that neutron irradiation can lead to increased Young's modulus of carbon fibers due to an increase in shear modulus of the graphite parallel to the basal planes [17]. The extent of deformation in different composite materials due to irradiation depends on the type of fibers, their reinforcement direction in the carbon matrix, and the interaction between the carbon fibers and the matrix [18]. Pitch-based carbon fibers have more resistance to degradation under exposure to irradiation than polyacrylonitrile-based carbon fibers. Low doses of neutron irradiation may result in the strengthening of graphite by ejecting atoms from graphene planes into interstitial positions, while at high doses, the ejection of atoms amounts to structural disorder and may account for strain-induced cracking which results in a decrease of the mechanical properties of carbon materials [19]. An investigation of the neutron irradiation effects on the mechanical properties of organic and carbon composite materials reported the improvement in flexural strength and modulus, while no effect on deformation at the break was observed, but the flexural strength was distinctly affected [15]. This investigation also obtained interesting information about the improvement in the ILSS of a composite system. However, neutron irradiation damages the matrix (particularly carbon), but it improves the matrix-fiber interaction. Induction of compressive stresses on the graphite fibers by a slight shrink of the matrix and/or swelling of the fibers can probably be attributed to the improvement of ILSS. When organic materials are subjected to neutron irradiation, the degradation is caused mainly by recoil particles that are produced by neutron reactions, such as elastic scattering, inelastic scattering, and nuclear reaction [20]. On the other hand, γ-ray irradiation induced degradations are primarily due to secondary electrons generated by the interaction with photons. Investigations have shown that as far as the E-glass and T-glass fabric composites are concerned, the dose dependence of the composite strength at 77 K is almost independent of the composite specifications, such as the kind of glass fibers, type of fabric weave, the specimen thickness, and the volume fraction of the reinforced fibers [21].

6.4 Effect of Atomic Oxygen

In the low Earth orbit, space environment atomic oxygen (AO) is the most prevalent chemical species. This presents at space station freedom altitudes between 300 and 500 km (150 nautical miles–270 nautical miles). AO is formed by molecular bond breakage of oxygen by solar photons at wavelengths below 0.243 μm [22]. Surfaces, which face the direction of travel of the spacecraft (the ram direction), are impacted with a high flux of atomic oxygen (approximately 10^{14} 10^{15} atoms/cm^{2}s) at collision energies of approximately 4.5 eV. A result of the impact between atomic oxygen and an organic polymer surface is fragmentation of the high molecular weight polymer chains, which leads to the formation of a volatile condensible material, which could deposit onto nearby surfaces resulting in contamination [23]. For example, oxidation products of silicones have been shown to produce brown contamination products on adjoining surfaces on the long duration exposure facility (LDEF) spacecraft. Also, results from the exposures of hydrocarbons to simulated atomic oxygen environments have shown evidence that gas phase reaction products of CO and CO_2 [24,25] are emitted, which could contaminate other spacecraft surfaces. Oxidation of polymer surfaces resulting in erosion and mass loss has been observed for materials aboard shuttle missions, such as STS-5 and STS-8 [26]. These chemical changes can lead to changes in surface morphology, strength, and thermal/optical properties.

6.5 Low Earth Orbit Space Environments

The low Earth orbit space environment constituents include high vacuum, UV radiation, thermal cycles, atomic oxygen, electromagnetic radiation, charged particles, micrometeoroids, and man-made debris [27]. These environmental constituents can significantly degrade the material characteristics of polymers and polymer matrix composite materials. The exposure of composite materials to a severe LEO space environment leads to the structural modification and mass loss by outgassing, thermal cycling can result in fatigue cracking, an AO attack leads to surface erosion, UV radiation can modify material properties and collisions with high velocity micrometeoroids, and man-made debris can lead to delamination in the laminated composites [27,28]. The outgassing of moisture and other volatile constituents can produce dimensional changes in composite components and also leads to contamination in adjacent spacecraft components [29,30]. A decrease in strength and stiffness of the graphite/epoxy composites after being exposed to thermal cycles was observed, and this reduction was found to be in exponential proportion with increasing thermal cycles [21]. The loss in transverse flexural strength and stiffness was quicker as compared to the other properties [22]. Matrix loss at the composite surface was considered as the prime cause of the severe drop in transverse flexural properties. The effect of hypervelocity impact on carbon/epoxy composites in a low Earth space environment is also reported by Awaja et al. [31].

References

1. T. Xu, G. Li, and S.-S. Pang, Effects of ultraviolet radiation on morphology and thermo-mechanical properties of shape memory polymer based syntactic foam, *Compos. Part A Appl. Sci. Manuf.*, 42(10): 1525–1533, 2011.
2. R. S. C. Woo, Y. Chen, H. Zhu, J. Li, J.-K. Kim, and C. K. Y. Leung, Environmental degradation of epoxy–organoclay nanocomposites due to UV exposure. Part I: Photo-degradation, *Compos. Sci. Technol.*, 67(15–16): 3448–3456, 2007.
3. J. E. Guillet, Fundamental processes in the UV degradation and stabilization of polymers, in *Chemical Transformations of Polymers*, R. Rado (Ed.). Oxford, UK: Butterworth-Heinemann, pp. 135–144, 1972.
4. A. Tcherbi-Narteh, M. Hosur, E. Triggs, and S. Jeelani, Thermal stability and degradation of diglycidyl ether of bisphenol A epoxy modified with different nanoclays exposed to UV radiation, *Polym. Degrad. Stab.*, 98(3): 759–770, 2013.
5. H. R. Phelps and J. E. R. Long, Property changes of a graphite/epoxy composite exposed to nonionizing space parameters, *J. Compos. Mater.*, 14(4): 334–341, 1980.
6. F. Larsson, The effect of ultraviolet light on mechanical properties of Kevlar 49 composites, *J. Reinf. Plast. Compos.*, 5(1): 19–22, 1986.
7. W. B. Liau and F. P. Tseng, The effect of long-term ultraviolet light irradiation on polymer matrix composites, *Polym. Compos.*, 19(4): 440–445, 1998.
8. H. Wang, H. Xie, Z. Hu, D. Wu, and P. Chen, The influence of UV radiation and moisture on the mechanical properties and micro-structure of single Kevlar fibre using optical methods, *Polym. Degrad. Stab.*, 97(9): 1755–1761, 2012.
9. R. Scaffaro, N. Tzankova Dintcheva, and F. P. La Mantia, A new equipment to measure the combined effects of humidity, temperature, mechanical stress and UV exposure on the creep behaviour of polymers, *Polym. Test.*, 27(1), 49–54, 2008.
10. F. Awaja and P. J. Pigram, Surface molecular characterisation of different epoxy resin composites subjected to UV accelerated degradation using XPS and ToF-SIMS, *Polym. Degrad. Stab.*, 94(4): 651–658, 2009.
11. Y. Gao, S. He, D. Yang, Y. Liu, and Z. Li, Effect of vacuum thermo-cycling on physical properties of unidirectional M40J/AG-80 composites, *Compos. Part B Eng.*, 36(4): 351–358, 2005.
12. S. Y. Park, H. S. Choi, W. J. Choi, and H. Kwon, Effect of vacuum thermal cyclic exposures on unidirectional carbon fiber/epoxy composites for low earth orbit space applications, *Compos. Part B Eng.*, 43(2): 726–738, 2012.
13. A. H. Baluch, Y. Park, and C. G. Kim, Hypervelocity impact on carbon/epoxy composites in low Earth orbit environment, *Compos. Struct.*, 96(Supplement C): 554–560, 2013.
14. J. W. McClure and W. J. Spry, Linear magnetoresistance in the quantum limit in graphite, *Phys. Rev.*, 165(3): 809–815, 1968.
15. S. Blazewicz et al., Effect of neutron irradiation on the mechanical properties of graphite fiber-based composites, *Carbon*, 40(5): 721–727, 2002.
16. R. E. Bullock and E. L. McKague, Radiation effects on mechanical properties of carbon/carbon composites, *Carbon*, 11(5): 547–553, 1973.
17. J.-B. Donnet and R. C. Bansal, *Carbon Fibers*. Boca Raton, FL: CRC Press, 1998.

18. K. Hamada, S. Sato, and A. Kohyama, Effects of neutron irradiation on microstructure and mechanical properties of carbon/carbon composites, *J. Nucl. Mater.*, 212–215(Part B): 1228–1233, 1994.

19. B. S. Brown, Radiation effects in superconducting fusion-magnet materials, *J. Nucl. Mater.*, 97(1): 1–14, 1981.

20. F. W. Clinard and G. F. Hurley, Ceramic and organic insulators for fusion applications, *J. Nucl. Mater.*, 103(Supplement C): 705–715, 1981.

21. J. W. T. Spinks and R. J. Woods, *An Introduction to Radiation Chemistry*, 3rd ed. New York: Wiley, 1990.

22. J. A. Dever, Low earth orbital atomic oxygen and ultraviolet radiation effects on polymers. Springfield, VA: NASA, 1991.

23. Y. Haruvy, Risk assessment of atomic-oxygen-effected surface erosion and induced outgassing of polymeric materials in LEO space systems, *ESA J. Eur. Space Agency*, 14(1): 109–119, 1990.

24. J. B. Cross, *Laboratory Investigations: Low Earth Orbit Environment Chemistry with Spacecraft Surfaces*. Los Alamos, NM: Los Alamos National Laboratory, 1989.

25. T. Nishikawa, K. Sonoda, and K. Nakanishi, Effect of atomic oxygen on polymers used as surface materials for spacecrafts, in *Proceedings of the Twenty-First Symposium on Electrical Insulating Materials*, Institute of Electrical Engineers of Japan, pp. 191–194, 1988.

26. J. T. Visentine and L. J. Leger, Material interactions with the low Earth orbital environment: Accurate reaction rate measurements, *JPL Publ.*, 87(14): 11–20, 1986.

27. S. Egusa and T. Seguchi, Polymer composites as magnet materials: Irradiation effects and degradation mechanism of mechanical properties, *J. Nucl. Mater.*, 179–181(Part 2): 1111–1114, 1991.

28. K.-B. Shin, C.-G. Kim, C.-S. Hong, and H.-H. Lee, Prediction of failure thermal cycles in graphite/epoxy composite materials under simulated low earth orbit environments, *Compos. Part B Eng.*, 31(3): 223–235, 2000.

29. J.-H. Han and C.-G. Kim, Low earth orbit space environment simulation and its effects on graphite/epoxy composites, *Compos. Struct.*, 72(2): 218–226, 2006.

30. D. K. Felbeck, Toughened graphite-epoxy composites exposed in near-Earth orbit for 5.8 years, *J. Spacecr. Rockets*, 32(2): 317–323, 1995.

31. F. Awaja, J. B. Moon, S. Zhang, M. Gilbert, C. G. Kim, and P. J. Pigram, Surface molecular degradation of 3D glass polymer composite under low earth orbit simulated space environment, *Polym. Degrad. Stab.*, 95(6): 987–996, 2010.

7

Loading Rate Sensitivity of Polymer Matrix Composites

7.1 Introduction

Massive demand of fiber-reinforced polymer (FRP) composite materials in various engineering and structural applications is primarily driven by their unique and tailorable properties, such as light weight, high specific strength and specific modulus, corrosion resistance, and good fatigue resistance. Diversified applications include various environmental and loading conditions, which range from quasi-static to dynamic/impact.

There are advanced applications of FRPs where they may deform rapidly. Such as when the composite jet engine compressor blades are exposed to the hazards of foreign object damage, bird impact on rotating blades. These impacts occur at velocities of up to 300 ms^{-1} and are capable of causing extensive damage to the blade material. Hence, it is essential to understand how the FRP composites respond under various ranges of loadings. Therefore, the reliable design of the composite components for impact resistance requires the detailed characteristics of the composite materials at high strain rates.

FRP composites are anisotropic materials, so the damages due to multi-axial stresses are complex phenomenon involving many failure mechanisms (frequently interactive) in micro- and macroscale. Over the years, several methods have been proposed and discussed, including fracture mechanics, nonlinear viscoplastic constitutive modeling, damage mechanics, and macroscopic (global) failure criteria. In the design and analysis of composite structures, which are mostly subjected to quasi-static loadings, the macroscopic (global) failure criteria based methods are the commonly adopted. Under certain biaxial states of stress and under dynamic loading conditions, available failure criteria are still not promising and fully reliable, as per the "worldwide failure exercises" [1]. These loadings are typically highly transient, and the material and structural response occurs over very short (dynamic) time scales (of the order of milliseconds or microseconds).

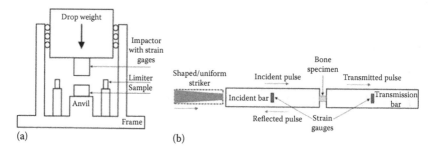

FIGURE 7.1

Schematic diagram of (a) drop tower impact tester (From Elibol, C. and Wagner, M. F.-X., *Mater. Sci. Eng. A*, 643, 194–202, 2015.) and (b) split Hopkinson pressure bar. (From Bekker, A. et al., *Mater. Sci. Eng. C*, 46, 443–449, 2015.)

There is a range of instruments available for loading rate/strain rate testing of FRPs. A servohydraulic testing machine can be utilized to generate quasi-static and low strain rates up to 10 s^{-1} (approximately) in the samples. A drop tower apparatus (shown in Figure 7.1a) can generate strain rates between 10 s^{-1} and approximately 200 s^{-1} and higher rates up to and exceeding 1,000 s^{-1} can be produced by means of a split Hopkinson pressure bar [or Kolsky bar]) [2–4]. A schematic diagram has been shown in Figure 7.1b.

The absorbed energy with increasing strain rate also increases up to 62.4%. This increase in energy absorption is beneficial in applications of composite structures under dynamic loading conditions. The design and analysis of glass/epoxy composite structures based on the mechanical properties obtained at lower crosshead stroke rates lead to a conservative design [7]. In dynamic loading, toughness and mechanical properties of composites are known to change, and this places limitations on their performance. A translaminar fracture could occur when there is through-thickness damage in glass-reinforced polymer (GRP) laminated composites [8]. Some researchers have given explicit empirical relations for the rate dependence of these mechanical properties [3,9,10]. High strain rate studies by Daniel et al. [11] and Gilat et al. [4] through uniaxial tensile test methods have shown considerable increases in stiffness and strength of FRP composites with increased strain rate.

7.2 Mechanical Properties of Fiber-Reinforced Polymer Composites in Tensile Loading under Different Strain Rates

A study on the unidirectional glass fiber/epoxy composites under tensile loading over a wide range of strain rates from 10^{-6} to 30 s^{-1} reported that the dynamic modulus is 50% higher than the static modulus and the dynamic strength is three times higher than the static strength [12]. However, for angle ply glass/epoxy laminates, Lifshitz [13] found that the elastic modulus and failure strain were insensitive to the strain rate and the failure stress in dynamic loading was only 20%–30% higher than the failure stress in static loading. Okoli and Smith [14,15] investigated the effects of strain rate on the tensile, shear, and flinvest properties of glass/ epoxy laminate in the range of speeds from 0.008 to 4 mm/s. Their results were in agreement with the results of the studies conducted by Armenakas and Sciamarella [16] at various strain rates (0.0265–30,000 min⁻¹) that reported a linear variation of the tensile modulus of elasticity of unidirectional glass fiber/epoxy composites with the log of strain rate. However, with the increase in strain rate, the ultimate tensile stress and strain of the composite decreased. Their study also indicated that there is a change in failure modes as the strain rate changes from quasi-static to dynamic, as shown in Figure 7.2.

Harding and Welsh [17,18] validated a dynamic tensile technique by performing tests (over the range of 10^{-4} to 1,000 s⁻¹) on glass/epoxy and carbon/ epoxy composites. The dynamic strength and modulus for the glass/epoxy composite were observed to be about twice the static value, whereas for carbon/epoxy, the failure stress, modulus, and failure mode of the composite were observed to be strain rate insensitive. Hayes and Adams [19] constructed a specialized pendulum impactor to study the strain rate sensitivity on the tensile properties of unidirectional glass/epoxy and carbon/epoxy composites. The strength and modulus of the glass/epoxy composites were observed to be rate insensitive at impact speeds in the range of 2.7–4.9 m/s, whereas the strength and modulus of carbon/epoxy composites decreased with increasing impact speed.

Daniel and Liber [20,21] also investigated the effect of strain rate (in the strain range of 10^{-4} to 27 s⁻¹) on the mechanical properties of unidirectional

(a)   (b)

FIGURE 7.2
(a) Brittle failure with fiber breakage (tested at 1.7×10^{-2} mm s⁻¹) and (b) fiber bunch pull out with signs of fiber matrix adhesion (tested at 10 mm s⁻¹). (From Okoli, O. I., *Compos. Struct.*, 54, 299–303, 2001.)

S-glass/epoxy and carbon/epoxy composites and found that the tensile modulus and failure strength of both the composites were rate insensitive. The behavior of unidirectional glass/epoxy composite materials at quasi-static (approximately 0.001 s^{-1}) and dynamic strain rates (from 1^{-1} to 100 s^{-1}) were investigated by Shokrieh and Omidi [22] using a servohydraulic testing apparatus. The experimental results show an increase in tensile modulus, strength, strain to failure, and absorbed failure energy of 12%, 52%, 10%, and 53%, respectively. For an increase in the loading rate from a static condition (0.0216 mm/s) up to a dynamic loading (1,270 mm/s), the dynamic strength of glass/epoxy composites increases 1.5 times with respect to the static strength, as shown in Figure 7.3.

Melin and Asp [23] investigated the strain rate dependence of the transverse tensile properties of a high performance carbon fiber/epoxy composite loaded in transverse tension. The specimens were tested under quasi-static and dynamic loading conditions (10^{-3} to 10^{3} s^{-1}). The initial transverse modulus was found to decrease slightly with increased strain rate, while the average transverse modulus was observed to be independent of strain rate. With

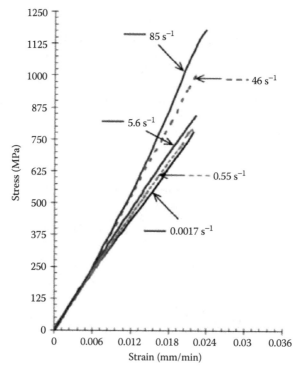

FIGURE 7.3
Typical stress–strain tensile behavior of glass/epoxy composites under various strain rates. (From Shokrieh, M. M. and Omidi, M. J., *Compos. Struct.*, 88, 595–601, 2009.)

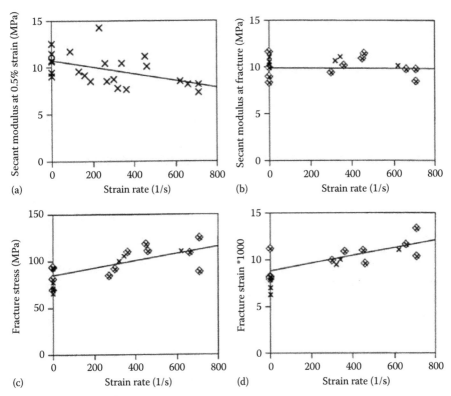

FIGURE 7.4

Mechanical properties for the specimen, plotted against strain rate at $\varepsilon = 0.5\%$. (a) Secant modulus at $\varepsilon = 0.5\%$. (b) Secant modulus just before failure. (c) Failure stress. (d) Failure strain. In (b)–(d) the specimens that failed at the edge of the gauge section where the radius begins are emphasized with a square. (From Melin, L. G. and Asp, L. E., *Compos. Part A Appl. Sci. Manuf.*, 30, 305–316, 1999.)

an increased strain rate, the stress and strain at failure were found to increase slightly. Thus, it was concluded that when the carbon/epoxy composite is loaded in the transverse direction it could exhibit a weak dependence on strain rate, as shown in Figure 7.4.

Daniel et al. [24] studied the dynamic tension response of unidirectional carbon/epoxy composites at high strain rates (up to 500 s^{-1}) using an internal pressure pulse generated explosively through a liquid medium. A tension test in longitudinal direction revealed that the modulus increased moderately with the strain rate (up to 20% over the static value), but the ultimate strain and strength did not vary significantly. The strength and modulus increased sharply over static values in the transverse direction and the slight increase in the ultimate strain was noticed. There was a 30% increase in the in-plane shear modulus and strength.

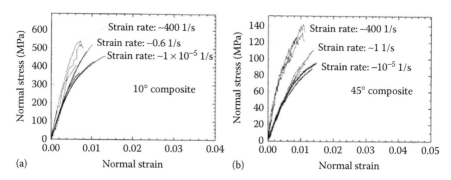

FIGURE 7.5
Stress–strain curves at different strain rates for (a) 10° and (b) 45° composites. (From Gilat, A. et al., *Compos. Sci. Technol.*, 62, 1469–1476, 2002.)

Gilat et al. [4] investigated the tensile behavior of carbon/epoxy composites, using a hydraulic testing machine for the quasi-static and intermediate tests and a tension split Hopkinson bar apparatus for the high strain rate tests. Tensile tests were performed for fiber orientations of 90°, 10°, 45° and [±45°]$_s$ at strain rates ranging from 10 s^{-5} to 650 s^{-1}. A significant increase in the stiffness was reported with increased strain rate in all of the configurations tested, as shown in Figure 7.5. For 45° and [±45°]$_s$, the layup configuration had a significant effect on the strain rate of the maximum stress, while for 90° and 10°, the layup configuration had a slight increase on the maximum stress, with an increased strain rate reported. Further, the maximum strain at all strain rates in the tests with the [±45°]$_s$ layups is much larger than in all the other types of test configurations.

Daniel et al. [11] conducted multiaxial experiments on a unidirectional carbon/epoxy material at three strain rates, quasi-static, intermediate, and high, 10^{-4}, 1, and 180–400 s^{-1}, respectively. A Hopkinson bar apparatus was used and specimens were loaded in an off-axis direction (to produce stress states combining transverse normal and in-plane shear stresses) [11]. The basic matrix-dominated mechanical properties of the composite, including the initial transverse and in-plane shear moduli, E_2 and G_{12}, the transverse tensile and compressive strengths, F_{2t} and F_{2c}, and the in-plane shear strength, F_6, that were derived from the transverse (90°) and off-axis stress–strain curves are shown in Table 7.1 [11].

Daniel et al. [25] proposed strain rate-dependent engineering failure criteria, which can be easily implemented in the design of composite structures undergoing small, nearly elastic dynamic deformations.

TABLE 7.1

Matrix Dominated Properties of Carbon/Epoxy Material (AS4/3501-6)

Properties	Strain Rate 0.0001 S⁻¹	Strain Rate 1 S⁻¹	Strain Rate 400 S⁻¹
Transverse modulus, E_2 (GPa)	11.2	12.9	14.5
Shear modulus, G_{12} (GPa)	7.0	8.0	9.0
Modulus ratio, $\alpha = (E_2/G_{12})$	1.60	1.57	1.61
Transverse tensile strength, F_{2t} (MPa)	65	[80]	[90]
Transverse compressive strength, F_{2c} (MPa)	285	345	390
Shear strength, F_6 (MPa)	80	[95]	[110]
Strength ratio, F_{2c}/F_6	3.56	3.63	3.55

Note: Numbers in brackets denote extrapolated values.

Compression dominated failure:

$$\left(\frac{\sigma_2^*}{F_{2c}}\right)^2 + \alpha^2 \left(\frac{\tau_6^*}{F_{2c}}\right)^2 = 1.$$

Shear dominated failure:

$$\left(\frac{\tau_6^*}{F_6}\right)^2 + \frac{2}{\alpha}\left(\frac{\sigma_2^*}{F_6}\right) = 1.$$

Tension dominated failure:

$$\frac{\sigma_2^*}{F_{2t}} + \frac{\alpha^2}{4}\left(\frac{\tau_6^*}{F_{2t}}\right)^2 = 1,$$

where $\alpha = E_2/G_{12}$ and

$$\sigma_i^* = \sigma_i \left(m_f \log \frac{\dot{\varepsilon}}{\dot{\varepsilon}_0} + 1 \right)^{-1}, \sigma_i = \sigma_2, \tau_6.$$

The symbols $m_f = 0.057$, $\tau_6 =$ shear stress, $\sigma_2 =$ uniaxial stress normal to the fiber direction, $\dot{\varepsilon}_0 =$ reference strain (10^{-4} for quasi-static loading), and $\dot{\varepsilon} =$ strain rate.

The experimental strengths were plotted with stresses obtained from classical failure criteria, (maximum stress, maximum strain, Tsai-Hill, and Tsai-Wu) and failure mode based and partially interactive criteria (Hashin-Rotem, Sun, and Daniel) [25–28], as shown in Figure 7.6.

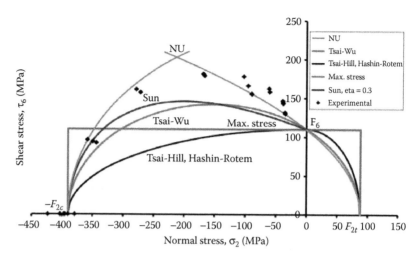

FIGURE 7.6
Comparison of theoretical failure envelopes and experimental results for AS4/3501-6 carbon/epoxy composite under high rate transverse normal and shear stress. (From Daniel, I. M. et al., *Compos. Sci. Technol.*, 71, 357–364, 2011.)

Inherent anisotropy in a carbon/epoxy composite between the through-thickness and in-plane properties resulted in interlaminar failures under interlaminar shear stress, τ_5, and transverse normal, σ_3. A comparison of experimental results of Daniel et al. [25] with various theories has been presented in Figure 7.7.

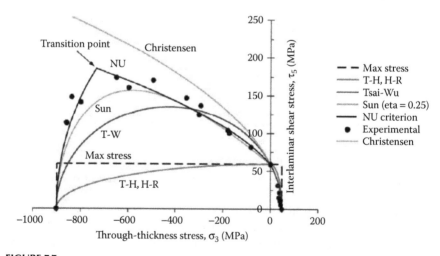

FIGURE 7.7
Failure envelopes for carbon-fabric/epoxy composite under normal through-thickness and interlaminar shear loading (comparison of experimental results [25] and predictions of various theories including NU theory [developed by Northwestern University (NU)]). (From Daniel, I. M., *Strain*, 43, 4–12, 2007.)

7.3 Mechanical Properties of Fiber-Reinforced Polymer Composites in Compressive Loading under Different Strain Rates

The dynamic compressive response of unidirectional and transversely isotropic glass/epoxy composites has been studied by Kumar et al. [29], using the Kolsky pressure bar technique for fiber orientations of 0°, 10°, 30°, 45°, 60°, and 90° at an average strain rate of 265 s⁻¹. The compressive behavior of glass fiber/epoxy composites was found to be strain rate sensitive for all fiber orientations, as shown in Figure 7.8. Compared to quasi-static (2×10^{-4} s⁻¹), the dynamic ultimate strength increased almost 100% for 0°, 80% for 10° fiber orientations, and about 45% for all other orientations. Composite specimens of 0° orientation fractured along the fibers by tensile splitting, this can be attributed to the formation of transverse tensile strains because of Poisson's effect under compressive loading. Specimens of 10°, 30°, and 45° fiber orientation fractured along the fiber mainly by interlaminar shear, although cracks that resulted by a degree of tensile splitting were also observed on the surface of some of the specimens. They also noticed that the dynamic stress–strain curves were linear up to the fracture for fiber orientations of 0° and 10° and nonlinear for orientations greater than 10°.

An investigation into the effect of strain rate (in the strain range, 5×10^{-4} s⁻¹ to 2,500 s⁻¹) on pure epoxy resin and cross-woven, glass-fiber-reinforced epoxy under compressive loading was studied by Tay et al. [30]. The experimental results on the response of pure epoxy and glass fiber-reinforced polymer (GFRP) revealed that both are strain rate sensitive, mainly in the low strain rate, as shown in Figure 7.9. A marked increase in the dynamic

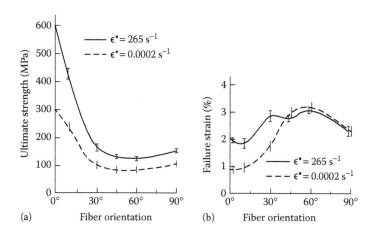

FIGURE 7.8
Variation of ultimate strength (a) and failure strain (b) with fiber orientation for unidirectional GFRP composite at different strain rates. (From Kumar, P. et al., *Mater. Lett.*, 4, 111–116, 1986.)

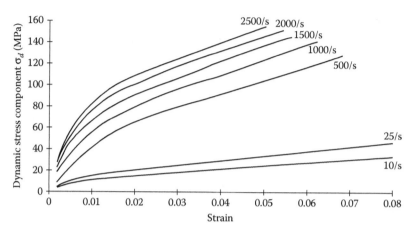

FIGURE 7.9
Dynamic stress component for GFRP. (From Tay, T. E. et al., *Compos. Struct.*, 33, 201–210, 1995.)

modulus was reported with increasing strain rate. It was observed that the stress–strain response under dynamic loading is a function of strain state and strain rate.

Lowe [31] studied strain rate effects on transverse mechanical properties of T300/914 carbon/epoxy unidirectional composites at various strain rates in transverse compression tests. The experimental results indicated an increase in both transverse modulus and compressive strength with an increasing strain rate. Vural and Ravichandran [32] studied the transverse failure behavior of thick unidirectional S2-glass fiber/epoxy composites at strain rates from 10^{-4} to $10^4 s^{-1}$. Their experimental results indicated that the compressive strength increased with the increment of the strain rate. Tsai and Kuo [33] studied the effect of strain rate from 10^{-4} to 500 s^{-1} on the transverse compressive strength of glass fiber/epoxy and carbon fiber/epoxy composites using a hydraulic material testing system (MTS) machine and a split Hopkinson pressure bar. For both composite systems, the transverse compressive strength was found to increase with increasing strain rates. Inspection of the compression failed specimens using the scanning electron microscope revealed that for the glass fiber/epoxy composites, the main failure mode was due to the matrix shear failure; however, for the carbon fiber/epoxy composites, it was the fiber/matrix interfacial debonding, which might dramatically reduce the transverse compressive strength of the composites. Using a high-speed servo-hydraulic machine by Shokrieh and Omidi [34], dynamic transverse lamina properties of unidirectional glass fiber/epoxy composites are extracted from tensile and compressive test results. For both the tensile and compressive loading cases, the obtained transverse lamina strength and modulus response show a clear strain rate dependency, as shown in Figure 7.10. For an increase

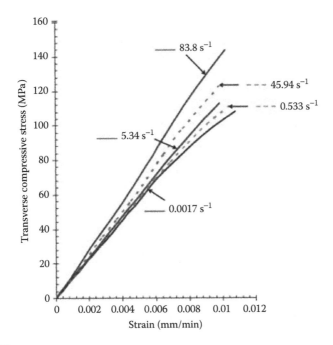

FIGURE 7.10
Transverse compressive stress–strain curves at different strain rates for glass/epoxy composites. (From Shokrieh, M. M. and Omidi, M. J., *Compos. Struct.*, 93, 690–696, 2011.)

in the strain rate from 0.001 s⁻¹ to 84 s⁻¹, there is an increase of 41.36% in tensile strength and 13.78% in tensile modulus. The corresponding values for compressive strength and modulus are about 31.37% and 23.36%, respectively. The transverse tensile failure strain shows an increase of 16% as the strain rate changes from quasi-static to dynamic, while the transverse compressive failure strain decreases with increased strain rate.

Hosur et al. [35] studied the dynamic response of unidirectional carbon/epoxy composites under transverse loading using a modified split Hopkinson pressure bar set up at three different strain rates of 82, 164, and 817 s⁻¹. Their experimental results reported a 25%–50% increase in the modulus and a 0.6%–25% increase in transverse strength under dynamic loading as compared to static values. Figure 7.11 shows the stiffening effect in the composite with the change in loading from quasi-static to dynamic. However, a decrease in stiffness was also observed with a further increase in strain rate. This behavior has been attributed to the viscoelastic nature of the polymer matrix and the change in failure mode with the change in strain rate. The dominating failure modes under in-plane compression loading were observed to be delamination splitting and crushing.

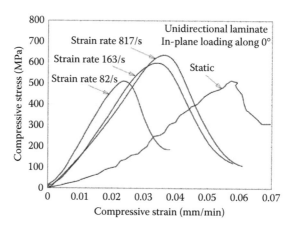

FIGURE 7.11

Compressive stress–strain variation at different strain rates for unidirectional glass/epoxy composite. (From Hosur, M. V. et al., *Compos. Struct.*, 52, 405–417, 2001.)

7.4 In-Plane Shear Behavior at Different Strain Rates

The results of the investigations by Harding and Welsh [17] on the −45° glass/epoxy composite and Staab and Gilat's [36] on the ±45° glass/epoxy composite specimens indicated a sensible increase of laminate strength with strain rate of the order of 1,000 s⁻¹. This increase in the laminate strength reflects a large increase in the shear strength. Al-Salehi et al. [37] obtained the lamina in-plane shear properties at various rates of strain on glass/ epoxy and Kevlar/epoxy filament wound tubes with winding angles ±55° and ±65°, for both, under internal hoop loading. The results obtained from ±55° specimens indicated that with increasing strain rate from 0 to 400 s⁻¹, the shear strength is increased by 70% for glass/epoxy and 115% for Kevlar/ epoxy materials. The results extracted from ±65° specimens were lower than the ±55° specimens. Tsai and Sun and Shokrieh and Omidi [38,39] studied the strain rate effect (up to 700 s⁻¹) on the in-plane shear strength of unidi- rectional off-axis S2/8552 glass fiber/epoxy laminate composites using a split Hopkinson pressure bar. The specimens were tested at fiber orientations of 15°, 30°, 45°, and 60°. They modeled stress–strain curves based on a visco- plasticity model established at the lower strain rate data. Further, this model is extended to high strain rates up to 700 s⁻¹ [38]. The shear strain rate was also extracted from the axial strain rate by relating the effective plastic strain rate to the plastic shear strain rate on the basis of a viscoplasticity model [39]. The experimental results showed that, in all cases, the shear strength of the glass fiber/epoxy composite was quite sensitive to strain rate and the shear strength increased as the strain rate increased.

With a high-speed servohydraulic test by Shokrieh and Omidi [39], in-plane shear failure properties of unidirectional glass/epoxy composites were studied at various stroke rates from 0.0216 to 1,270 mm/s. The dynamic shear strength response showed an increase of approximately 37% over the measured quasi-static value. The shear modulus and shear strain to failure of the composite decreased with the increase in strain rate (Figure 7.12).

Daniel et al. [40] investigated the strain rate sensitivity on in-plane shear properties of carbon/epoxy composites up to 500 s^{-1} strain rates. The results indicated that the in-plane dynamic shear strength and shear modulus increased approximately 30% over static values, while the dynamic ultimate shear strain was lower than the static one. Raju et al. [41] studied experimentally the in-plane shear responses of carbon fabric/epoxy and glass fiber/epoxy composites using a servohydraulic testing machine at nominal cross-head velocities ranging between 2.5×10^{-5} and 12.7 m/s. The V-notch rail shear specimen configuration was used for characterizing the in-plane shear properties of the composite systems. During the tests, a maximum estimated shear strain rate of 500 s^{-1} was achieved up to shear strain levels of 0.08 radians. The experimental results reported that at the highest strain rate, the shear strengths increased by a factor of three relative to that of the quasi-static rate and were independent of the reinforcement type.

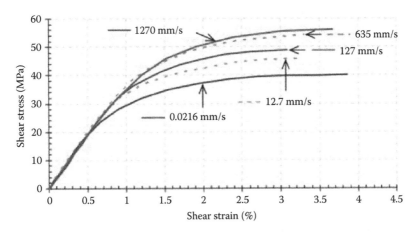

FIGURE 7.12
Typical in-plane shear stress–strain response of glass/epoxy composites under various stroke rates. (From Shokrieh, M. M. and Omidi, M. J., *Compos. Struct.*, 91, 95–102, 2009.)

7.5 Loading Rate Sensitivity of Environmentally Conditioned Fiber-Reinforced Polymer Composites

Ray [42] assessed the loading rate sensitivity of hygrothermally conditioned E-glass/epoxy and E-glass/unsaturated polyester composites. For both the systems, the interlaminar shear strengths (determined by short beam shear test) were higher at the higher loading rate (that is, 50 mm/min).

Ray [44] studied the loading rate sensitivity of ultra-low temperature-conditioned (−40°C, −60°C, and −80°C temperatures) E-glass fiber/epoxy composites with 55, 60, and 65 weight percentages. It was reported that the loading rate sensitivity of the polymer composites appeared to be inconsistent and contradictory at some points of the conditioning time, as well as at a temperature of conditioning. The phenomena may be attributed to low-temperature hardening, matrix cracking, and misfit strains. The loading rate sensitivity of freeze thaw-conditioned glass fiber/polyester composites is also investigated [45]. Loading rate sensitivity is strongly evident at a lower range of crosshead speed (0.5–50 mm/min) and interlaminar shear strength (ILSS) values are found to increase in all situations with more loading speed in the range. Thereafter, the fall in ILSS value is observed with higher crosshead speed. The ILSS of thermal shock-conditioned glass fiber/epoxy composites has also indicated the loading rate sensitivity when tested in a short beam shear test at two different loading rates, viz 2 and 10 mm/min [46]. The ILSS values were higher at 10 mm/min. The investigation on hygrothermally conditioned glass fiber/epoxy and glass fiber/polyester composite systems revealed that the ILSS of both the composite systems is strain-rate sensitive, as shown in Figure 7.13 [42,43]. The strain rate sensitivity is less pronounced

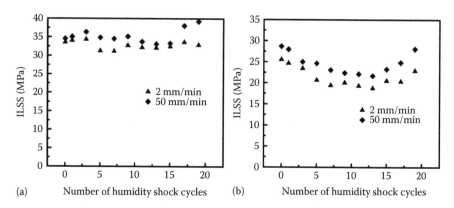

(a)   Number of humidity shock cycles   (b)   Number of humidity shock cycles

FIGURE 7.13

Variation of the ILSS with the number of humidity shock (at constant temperature) cycles at 2 mm/min (▲) and 50 mm/min (♦) crosshead speeds for (a) glass fiber-reinforced epoxy and (b) glass fiber-reinforced polyester composites. (From Ray, B. C., *J. Reinf. Plast. Compos.*, 25, 1227–1240, 2006; Ray, B. C., *Mater. Sci. Eng. A*, 379, 39–44, 2004.)

at higher conditioning times [43]. The freezing of absorbed moisture inside the composite leads to further damaging effects. These degradative effects of further freezing treatment are more evident at a lower loading speed. The state of the fiber/matrix interface after hygrothermal ageing may introduce more complications in evaluating the loading rate sensitivity of fiber-reinforced composites. The effect of changing seawater temperature during immersion ageing of glass/epoxy and glass/polyester composites on ILSS has been shown by short beam shear test at two crosshead velocities, viz 2 and 50 mm/min [47]. The shear strength values obtained were higher at all points of the cyclic environment at higher crosshead speeds.

The variation of ILSS for a glass/epoxy composite system *in-situ* conditioned at +50°C, +100°C, −50°C, and −100°C temperatures and tested at 1, 100, 200, 500, 700, and 1,000 mm/min loading rates, is shown in Figure 7.14a [48]. It is clearly evident from the figure that the above-ambient and subambient temperature exposures alter the ILSS values and, further, at each temperature, the ILSS has the loading rate-sensitive phenomenon, the ILSS decreased. The maximum ILSS for the glass/epoxy composite was obtained at −100°C temperature and 1 mm/min loading speed, with an increase of 85.72% more than the ILSS value obtained at ambient temperature at 1 mm/min. The greater value of the ILSS at a low loading speed can be attributed to a longer relaxation time, resulting in improved interfacial integrity of the composite material. Higher crosshead speed during testing minimizes the relaxation process

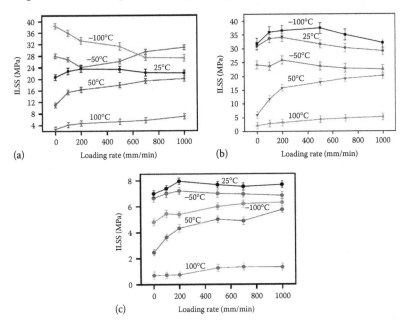

FIGURE 7.14
Variation of ILSS with loading rate for (a) glass fiber/epoxy, (b) carbon fiber/epoxy, and (c) Kevlar fiber/epoxy composites.

at the crack tip. This could be the reason for reduced ILSS values at higher crosshead speed. At a −50°C temperature, initially, the ILSS decreases from 1 to 200 mm/min and then increased with a further increase in the loading rate. The slight fall in the value at 200 mm/min conditioning could be related to the lower degree of cryogenic compressive stresses at the fiber/matrix interface. At +50°C and +100°C temperatures, the ILSS increased with an increasing loading rate. The reason may be the induced thermal stresses in the matrix region. Thermal stress induced microcracks in the polymer matrix and/or at the fiber/ matrix interface may possibly grow without blunting at a steady state. Some microcracks turn to potential cracks at low loading rates and cause significant reduction in interlaminar shear strength of the composite system, while as the loading rate increases, the time available to propagate the microcracks is less. This can be attributed to higher ILSS at higher loading rates at these above-ambient temperatures. The effects of microcracks and fiber breakage can nucleate the other form of damage, such as delamination, hence degradation in the thermomechanical properties of the composite occurred [49].

It was readily observed that at a −100°C temperature, the carbon fiber/ epoxy composites possess better ILSS compared to other testing temperatures. At a −100°C temperature, the variation of ILSS with a loading rate is shown in Figure 7.14b. It can be seen from this figure that as the rate of loading increases, the ILSS of the composite also increases up to 500 mm/min, but, after 500 mm/min, shear values decrease. However, at high temperatures of +50°C and +100°C, there is a significant change in ILSS values with loading speed. These results are probably due to the shear band propagation in the matrix resin at high temperatures. At low loading rate the propensity of crack propagation in the resin matrix is relatively higher due to the crack jumping (unstable or stick-slip) mode.

Kevlar fiber/epoxy composites exhibited higher ILSS at ambient temperatures as compared to above-ambient and subambient temperatures, as shown in Figure 7.14c. The variation of ILSS here is the net result of adhesion at the interface by physical and mechanical bonding at the interface. The thermal conditioning is likely to change the chemistry at the fiber/matrix interface. The unique chemistry and morphology of Kevlar fiber is also manifested by the composite behavior [50]. The bond between the Kevlar fiber and the surrounding matrix can be weakened by exposure to active environments.

References

1. M. J. Hinton, A. S. Kaddour, and P. D. Soden, Chapter 1.1: The world-wide failure exercise—Its origin, concept and content, in *Failure Criteria in Fibre-Reinforced-Polymer Composites*, Oxford, UK: Elsevier, pp. 2–28, 2004.
2. H. M. Hsiao and I. M. Daniel, Strain rate behavior of composite materials, *Compos. Part B Eng.*, 29(5): 521–533, 1998.

3. H. M. Hsiao, I. M. Daniel, and R. D. Cordes, Strain rate effects on the transverse compressive and shear behavior of unidirectional composites, *J. Compos. Mater.*, 33(17): 1620–1642, 1999.
4. A. Gilat, R. K. Goldberg, and G. D. Roberts, Experimental study of strain-rate-dependent behavior of carbon/epoxy composite, *Compos. Sci. Technol.*, 62(10): 1469–1476, 2002.
5. C. Elibol and M. F.-X. Wagner, Strain rate effects on the localization of the stress-induced martensitic transformation in pseudoelastic NiTi under uniaxial tension, compression and compression–shear, *Mater. Sci. Eng. A*, 643(Supplement C): 194–202, 2015.
6. A. Bekker, T. J. Cloete, A. Chinsamy-Turan, G. N. Nurick, and S. Kok, Constant strain rate compression of bovine cortical bone on the split-Hopkinson pressure bar, *Mater. Sci. Eng. C*, 46(Supplement C): 443–449, 2015.
7. M. M. Shokrieh and M. J. Omidi, Compressive response of glass–fiber reinforced polymeric composites to increasing compressive strain rates, *Compos. Struct.*, 89(4): 517–523, 2009.
8. M. Z. Shah Khan, G. Simpson, and E. P. Gellert, Resistance of glass-fibre reinforced polymer composites to increasing compressive strain rates and loading rates, *Compos. Part A Appl. Sci. Manuf.*, 31(1): 57–67, 2000.
9. J. R. Vinson and E. Woldesenbet, Fiber orientation effects on high strain rate properties of graphite/epoxy composites, *J. Compos. Mater.*, 35(6): 509–521, 2001.
10. N. K. Naik and V. R. Kavala, High strain rate behavior of woven fabric composites under compressive loading, *Mater. Sci. Eng. A*, 474(1): 301–311, 2008.
11. I. M. Daniel, B. T. Werner, and J. S. Fenner, Strain-rate-dependent failure criteria for composites, *Compos. Sci. Technol.*, 71(3): 357–364, 2011.
12. A. Rotem and J. M. Lifshitz, Longitudinal strength of unidirectional fibrous composite under high rate of loading, in *Proceedings of the 26th Annual Technical Conference of Society of the Plastics Industry Reinforced Plastics, Composites Division*, Washington, DC, pp. 1–10, 1971.
13. J. M. Lifshitz, Impact strength of angle ply fiber reinforced materials, *J. Compos. Mater.*, 10(1): 92–101, 1976.
14. O. I. Okoli and G. F. Smith, Overcoming inertial problems in the high strain rate testing of a glass/epoxy composite, *ANTEC'95*, 2: 2998–3002, 1995.
15. O. I. Okoli, The effects of strain rate and failure modes on the failure energy of fibre reinforced composites, *Compos. Struct.*, 54(2–3): 299–303, 2001.
16. A. E. Armenakas and C. A. Sciammarella, Response of glass-fiber-reinforced epoxy specimens to high rates of tensile loading, *Exp. Mech.*, 13(10): 433–440, 1973.
17. J. Harding and L. M. Welsh, A tensile testing technique for fibre-reinforced composites at impact rates of strain, *J. Mater. Sci.*, 18(6): 1810–1826, 1983.
18. L. M. Welsh and J. Harding, Effect of strain rate on the tensile failure of woven reinforced polyester resin composites, *J. Phys. Colloq.*, 46(C5): C5-405–C5-414, 1985.
19. S. V. Hayes and D. F. Adams, Rate sensitive tensile impact properties of fully and partially loaded unidirectional composites, *J. Test. Eval.*, 10(2): 61–68, 1982.
20. I. M. Daniel and T. Liber, Strain rate effects on mechanical properties of fiber composites, Part 3. Chicago, IL: Illinois Institute of Technology, 1976.
21. I. M. Daniel and T. Liber, Testing of fiber composites at high strain rates, in *Proceedings of the 2nd International Conference on Composite Materials, ICCM II*, Toronto, Canada, 1978.

22. M. M. Shokrieh and M. J. Omidi, Tension behavior of unidirectional glass/epoxy composites under different strain rates, *Compos. Struct.*, 88(4): 595–601, 2009.

23. L. G. Melin and L. E. Asp, Effects of strain rate on transverse tension properties of a carbon/epoxy composite: Studied by moiré photography, *Compos. Part A Appl. Sci. Manuf.*, 30(3): 305–316, 1999.

24. I. M. Daniel, H. M. Hsiao, and R. D. Cordes, Dynamic response of carbon/epoxy composites, *High Strain Rate Eff. Polym. Met. Ceram. Matrix Compos. Adv. Mater.*, 48: 167–177, 1995.

25. I. M. Daniel, J.-J. Luo, P. M. Schubel, and B. T. Werner, Interfiber/interlaminar failure of composites under multi-axial states of stress, *Compos. Sci. Technol.*, 69(6): 764–771, 2009.

26. C. T. Sun, 1.20 - Strength analysis of unidirectional composites and laminates, in *Comprehensive Composite Materials*, A. Kelly and C. Zweben (Eds.). Oxford, UK: Pergamon, pp. 641–666, 2000.

27. I. M. Daniel, Failure of composite materials, *Strain*, 43(1): 4–12, 2007.

28. I. M. Daniel, J.-J. Luo, and P. M. Schubel, Three-dimensional characterization of textile composites, *Compos. Part B Eng.*, 39(1): 13–19, 2008.

29. P. Kumar, A. Garg, and B. D. Agarwal, Dynamic compressive behaviour of unidirectional GFRP for various fibre orientations, *Mater. Lett.*, 4(2): 111–116, 1986.

30. T. E. Tay, H. G. Ang, and V. P. W. Shim, An empirical strain rate-dependent constitutive relationship for glass-fibre reinforced epoxy and pure epoxy, *Compos. Struct.*, 33(4): 201–210, 1995.

31. A. Lowe, Transverse compressive testing of T300/914, *J. Mater. Sci.*, 31(4): 1005–1011, 1996.

32. M. Vural and G. Ravichandran, Transverse failure in thick S2-glass/epoxy fiber-reinforced composites, *J. Compos. Mater.*, 38(7): 609–623, 2004.

33. J. L. Tsai and J. C. Kuo, Investigating strain rate effect on transverse compressive strength of fiber composites, *Key Eng. Mater.*, 306: 733–738, 2006.

34. M. M. Shokrieh and M. J. Omidi, Investigating the transverse behavior of glass-epoxy composites under intermediate strain rates, *Compos. Struct.*, 93(2): 690–696, 2011.

35. M. V. Hosur, J. Alexander, U. K. Vaidya, and S. Jeelani, High strain rate compression response of carbon/epoxy laminate composites, *Compos. Struct.*, 52(3): 405–417, 2001.

36. G. H. Staab and A. Gilat, High strain rate response of angle-ply glass/epoxy laminates, *J. Compos. Mater.*, 29(10): 1308–1320, 1995.

37. F. A. R. Al-Salehi, S. T. S. Al Hassani, N. M. Bastaki, and M. J. Hinton, Derived dynamic ply properties from test data on angle ply laminates, *Appl. Compos. Mater.*, 4(3): 157–172, 1997.

38. J.-L. Tsai and C. T. Sun, Strain rate effect on in-plane shear strength of unidirectional polymeric composites, *Compos. Sci. Technol.*, 65(13): 1941–1947, 2005.

39. M. M. Shokrieh and M. J. Omidi, Investigation of strain rate effects on in-plane shear properties of glass/epoxy composites, *Compos. Struct.*, 91(1): 95–102, 2009.

40. I. M. Daniel, R. H. LaBedz, and T. Liber, New method for testing composites at very high strain rates, *Exp. Mech.*, 21(2): 71–77, 1981.

41. K. S. Raju, C. K. Thorbole, and S. Dandayudhapani, Characterization of in-plane shear properties of laminated composites at high strain rates, in *Collection of Technical Papers AIAA/ASME/ASCE/AHS/ASC Structures, Structural Dynamics and Materials Conference*, pp. 7877–7884, 2006.

42. B. C. Ray, Effects of changing environment and loading speed on mechanical behavior of FRP composites, *J. Reinf. Plast. Compos.*, 25(12): 1227–1240, 2006.

43. B. C. Ray, Effects of crosshead velocity and sub-zero temperature on mechanical behaviour of hygrothermally conditioned glass fibre reinforced epoxy composites, *Mater. Sci. Eng. A*, 379(1–2): 39–44, 2004.

44. B. C. Ray, Loading rate effects on mechanical properties of polymer composites at ultralow temperatures, *J. Appl. Polym. Sci.*, 100(3): 2289–2292, 2006.

45. B. C. Ray, Freeze–thaw response of glass–polyester composites at different loading rates, *J. Reinf. Plast. Compos.*, 24(16): 1771–1776, 2005.

46. B. C. Ray, Thermal shock on interfacial adhesion of thermally conditioned glass fiber/epoxy composites, *Mater. Lett.*, 58(16): 2175–2177, 2004.

47. B. C. Ray, Effects of changing seawater temperature on mechanical properties of GRP composites, *Polym. Polym. Compos.*, 15(1): 59–63, 2007.

48. S. Sethi, D. K. Rathore, and B. C. Ray, Effects of temperature and loading speed on interface-dominated strength in fibre/polymer composites: An evaluation for in-situ environment, *Mater. Des.*, 65: 617–626, 2015.

49. A. Kelly and C. H. Zweben, *Comprehensive Composite Materials*. Amsterdam, the Netherlands: Elsevier, 2000.

50. B. C. Ray, Adhesion of glass/epoxy composites influenced by thermal and cryogenic environments, *J. Appl. Polym. Sci.*, 102(2): 1943–1949, 2006.

8

Environmental Durability of Fiber-Reinforced Polymer Nanocomposites

8.1 Introduction

The widespread acceptance of polymer composite materials is driven from their unique material properties, such as good impact strength, fatigue durability, improved corrosion resistance, superior specific strength, and modulus. It is well established that the excellent in-plane mechanical performance of laminated composite to a large extent is contributed by the fibrous phase. However, the properties measured during out-of-plane loading are very often limited and directly influenced by the matrix and the interface/interphase [1,2]. Hence, suitable modification of the matrix and/or interface is the key for applications involving out-of-plane mechanical loading. Earlier investigations have indicated the successful reinforcement of the polymer matrix by suitable micron-sized filler to enhance the strength and/or toughness of the polymer. However, the type of reinforcement, its size, shape, content, and distribution combined affect the performance of the reinforced polymer [3]. In conjunction with this, the interaction of the reinforcement filler with the polymer matrix and the stress transfer efficiency of the generated reinforcement/polymer interface are also decisive in this regard. The toughness enhancement in the case of the polymer matrix reinforced with micron-sized rigid glass spheres is attributed to the matrix shear yielding and crack pinning mechanisms [4]. In addition to these mechanisms, crack deflection and microcracking also contribute toward the improved toughness of the polymeric composites reinforced with second phase particles [5–8]. Soft rubbery particles also act as potential toughening agents in the brittle polymer, but this negatively affects the stiffness and thermal stability of the material.

The next-generation polymer composites are comprised of nanofillers having enormous high specific surface area. This helps in having large interfacial area for efficient stress transfer from the soft matrix to the strong reinforcement.

Researchers round the globe have successfully developed scientific techniques to improve the thermal, electrical, and mechanical properties of polymeric materials by suitable nanofiller reinforcement. However, the chemical structure of the nanofiller, its subsequent interaction with the polymer, and its controlled alteration by interface tailoring are some of the key factors, which combined decide the durability and reliability of these advanced materials [9]. Further, the character and chemistry of the constituents and interfaces of these polymer composites on interaction with the in-service environmental parameters reframe their long-term performance. Due to the in-built complexity of these multiscale composites, understanding must be initially gathered for the durability of nanofiller/polymer composites under various practical environments and then its logical extrapolation to nanofiller embedded fiber-reinforced polymer (FRP) composites. However, there is another possible way of nanoscale modification of FRP composites, that is, by fiber modification technique. The first objective in this type of modification is to deposit nanofillers on the fiber surface. This nanomodified fiber then can be used along with polymer matrix to fabricate nanophased FRP composites. The presence of nanofillers in the polymer matrix is expected to restrict the rate of crack propagation due to its inherent high strength and, thus, a substantial estimated improvement in the fracture toughness. Failure modes like diffused interlaminar and intralaminar damages may also be affected by the nanofiller reinforcement of the polymer.

This chapter starts with some basic fundamental concepts of polymer nanocomposites. The focus has been imparted on carbon-nanotube (CNT)-based composites due to its extensive acceptance as a potential nanofiller to strengthen and toughen the polymeric material. Moving to the environmental durability of these nanocomposites, the effects of various environmental parameters, namely, temperature, humidity, ultraviolet (UV) light and other high-energy irradiation have been discussed afterward.

8.2 Reinforcement Effect of Carbon Nanotube in Polymeric Materials

8.2.1 Why Nanofiller Reinforcement?

Effects of nanofiller size [10], shape [11], and volume content [12] on the final nanofiller-reinforced composites have been verified both theoretically and experimentally. One of the critical parameters in conventional nanofillers is the specific surface area (surface area per unit mass or volume), which is a function of its aspect ratio (length-to-diameter [l/d] ratio), which has a significant contribution toward the effective load transfer at the reinforcement/matrix interface. Figure 8.1 shows the surface/volume of nanofillers as a function of its diameter for spherical and cylindrical (with $l/d = 1$ and 10,000) shapes in a log–log scale. There is a drastic decrease in available surface area

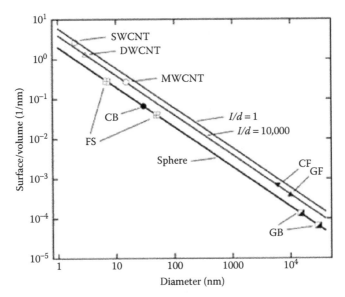

FIGURE 8.1
Ratio of surface area to volume for spherical and fibrous fillers as a function of the filler diameter. (From Fiedler, B. et al., *Compos. Sci. Technol.*, 66, 3115–3125, 2006.)

per unit volume with an increase in the particle diameter, which essentially gives the uniqueness of the particles in their nano dimensions. With a smaller particle size, more surface/interface area is available, which will facilitate more energy/stress transfer across the interface. Hence, even if glass fiber and carbon fiber have a similar aspect ratio as CNTs, still the CNTs exhibit extremely high surface/volume due to their sufficiently low diameter.

When the shape of the nanofillers is considered, usually the spherical fillers like glass balls, fused silica, and carbon black, and so on possess lower surface/volume in comparison to cylindrical fillers like CNTs, which enable CNTs to be a more appropriate candidate for having enormous interfacial area.

Another important factor in the case of nanofiller-reinforced polymer composites is the interparticle separation (i.e., the distance between the surfaces of two adjacent particles), which is a function of the size, shape, volume content, and distribution of the nanofiller. In the case of spherical nanoparticles with diameter "d" and volume fraction "v," the most uniform distribution resembles the face-centered cubic atomic ordering, where the interparticle separation (s_{fcc}^{sphere}) is determined by the following equation:

$$s_{fcc}^{sphere} = d \left[\frac{1}{2} \left(\frac{4\sqrt{2}\,\pi}{3v} \right)^{\frac{1}{3}} - 1 \right]$$

(8.1)

With the same spherical particle size and volume fraction, for a random arrangement, the interparticle separation (s_{ran}^{sphere}) becomes [14]

$$s_{ran}^{sphere} = \frac{2d(1-v)}{3v}$$

(8.2)

In a unidirectional fiber-reinforced composite, the ideal stacking arrangements are hexagonal and square packing. The interfiber surface separation distance in the case of hexagonal (s_{hex}^{fiber}) and square (s_{sq}^{fiber}) arrangements, with fiber diameter "d" and volume fraction "v" are given below [15]:

$$s_{hex}^{fiber} = d\left[\left(\frac{\pi}{2v\sqrt{3}}\right)^{\frac{1}{2}} - 1\right]$$

(8.3)

$$s_{sq}^{fiber} = d\left[\left(\frac{\pi}{4v}\right)^{\frac{1}{2}} - 1\right]$$

(8.4)

The above-mentioned equations for interfiber surface separation will be valid in the case of a CNT-reinforced composite when the packing is quite dense and with a certain directional alignment. The interfiller separation for spherical nanoparticles and fibers (CNT) is shown in Figure 8.2, as a function of the volume fraction of the filler. Interfiller separation has also a direct

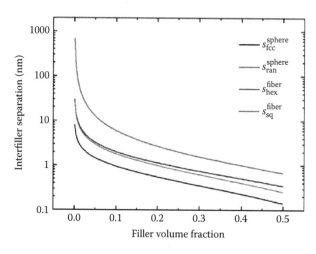

FIGURE 8.2

Effect of filler volume fraction, shape, and arrangement on the interfiller separation (all fillers are of 1 nm diameter).

influence on the fracture toughness of the material by a crack pinning mechanism, which has been described later.

It is very obvious that irrespective of the size and shape of the nanofiller, the interfiller separation decreases as the volume fraction of the nanofiller increases. It is very evident from the above-mentioned equations that with the same volume fraction of the filler, but with a large filler size, the interfiller separation increases. The maximum filler content in the composite is limited by this interfiller separation. Hence, with a higher volume fraction of the nanofiller, the interfiller separation decreases and eventually there is a significant chance for agglomeration.

The length of the CNT (l) plays a very significant role in deciding the load carrying capacity of the composite material. For an aligned CNT/polymer composite with CNT length less than the critical length [16], the longitudinal tensile strength of the composite material σ_c can be predicted from the below-mentioned equation [17]:

$$\sigma_c = \frac{l\tau_c}{d}v + (1-v)\sigma'_m \tag{8.5}$$

where:

τ_c is the smaller of the matrix shear yield strength or the fiber-matrix interfacial bond strength

d is the diameter of CNT and v is the volume fraction of CNT

σ'_m is the stress in the matrix when the composite fails

Hence, the shear strength of the CNT/polymer interface has a more pronounced effect than the CNT strength in the case of a CNT/polymer composite.

In the case of a thermoplastic polymer composite, the presence of nanofillers even in a very small quantity exhibit a very high surface area, which is sufficient for the nucleation of polymer crystals. In a CNT/poly(vinyl alcohol) (PVA) composite, the interfacial crystalline PVA (at the CNT/PVA interface) has a strong bonding with the CNT, resulting in an efficient stress transfer across the interface and, hence, fracture takes place at the interface between strong crystalline PVA and weak amorphous PVA. For a CNT/polymer composite with a crystalline coating thickness b (thickness of the CNT/polymer interface/interphase), the strength of the composite is predicted from the following equation [18]:

$$\sigma_c = \left(1 + \frac{2b}{d}\right)\left[\frac{l}{d}\tau_m - \left(1 + \frac{2b}{d}\right)\sigma_m\right]v + \sigma_m \tag{8.6}$$

where τ_m and σ_m are the shear strength and tensile strength of the polymer matrix, respectively.

Various approaches have also been adopted to predict the modulus of the short fiber-reinforced composite materials. For an aligned short fiber composite, the modulus (E_c) can be estimated from the following equation given by Cox [19]:

$$E_c = \left(n_l E_f - E_m\right)v + E_m \tag{8.7}$$

where:

E_f and E_m represent the modulus of fiber and matrix, respectively
n_l represents the length efficiency factor given by

$$n_l = 1 - \frac{\tanh\left(a\dfrac{l}{d}\right)}{a\dfrac{l}{d}}$$

and

$$a = \sqrt{\frac{-3E_m}{2E_f \ln v}}$$

The above-mentioned modulus expression represents that fibers (CNT) with a higher aspect ratio (l/d) are efficient for modulus enhancement.

For a nonaligned fiber composite, the modulus can be expressed as follows:

$$E_c = \left(n_o n_l E_f - E_m\right)v + E_m \tag{8.8}$$

The orientation factor n_o is "1" for perfectly aligned fibers and "1/5" for randomly oriented fibers [20,21].

Another model formulated by Halpin and Tsai, known as the "Halpin-Tsai" equation [22], predicts the modulus of fiber reinforced composite materials. For an aligned short fiber-reinforced composite, the composite modulus (E_c) is given by

$$E_c = E_m \left(\frac{1 + \zeta n v}{1 - n v}\right) \tag{8.9}$$

where $\zeta = 2\dfrac{l}{d}$ and $n = \dfrac{E_f - E_m}{E_f + E_m}$

For a randomly oriented fiber-reinforced composite, the composite modulus is expressed as follows:

$$E_c = E_m \left[\frac{3}{8}\left(\frac{1 + \zeta n_L v}{1 - n_L v}\right) + \frac{5}{8}\left(\frac{1 + 2n_T v}{1 - n_T v}\right)\right] \tag{8.10}$$

where $n_L = \dfrac{E_f - E_m}{E_f + \zeta E_m}$ and $n_T = \dfrac{E_f - E_m}{E_f + 2E_m}$

Hence, from all the above-mentioned theoretical expressions, we can conclude that CNT with a lower diameter, higher length, higher strength, and modulus promotes higher strength and stiffness enhancement in the resulting composites. In addition to that, the CNT/polymer interfacial strength and thickness also have important roles for effective and potential reinforcement.

8.2.2 Nanofiller/Polymer Interaction

Depending on the nature of a CNT/polymer interaction, the interfacial bonding may be divided into two types: (1) covalent and (2) noncovalent. The dominating type of bonding is decided depending on the physical, mechanical, and chemical affinity of a CNT and polymer toward each other. Bridging, wrapping, and interfacial interaction areas are some of the noncovalent bonding methods. When a single polymer chain interacts with multiple reinforcements, it is called bridging. The ratio between the radius of gyration of a polymer to the interfiller separation gives the probability of bridging [23]. Hence, a higher molecular weight of a polymer and a lower interfiller separation is advantageous to have effective bridging. A specific interaction area is defined as the total interfacial area per unit volume of the nanocomposite, which depends on the specific surface area and volume fraction of reinforcement and the densities of the constituents. A CNT with a lower diameter is thus preferable to have a higher specific interaction area.

Improved interaction and uniform distribution of nanofillers result due to wrapping [24,25]. The extent of wrapping is strongly affected by various characteristics of the polymer that include the stiffness of the backbone and chemical structure and size and shape of the nanofiller. When a polymer with a rigid backbone tends to wrap around a CNT, a helical configuration is achieved [26,27]. However, flexible polymers with bulky aliphatic or aromatic side groups, such as polystyrene and poly(methyl methacrylate) (PMMA), tend to form interchain coiling rather than a helical configuration [28]. A polymer with an aromatic group in its backbone has an affinity towards CNT, due to p–p interaction between them, which decides the adsorption conformation. However, such interaction is restricted due to the presence of side groups containing aliphatic chains. Geometrical aspects, such as polymer chain length and CNT diameter, also play an important role toward the extent of wrapping. A smaller diameter of the CNT with respect to the radius of gyration of the polymeric chain infers suitable interaction between the polymer and the CNT [23]. In this line, a higher extent of wrapping essentially requires a lower CNT diameter and a polymer with higher molecular weight.

The chemical structure of the monomeric unit of the polymer also plays an important role in the interfacial bond strength of the CNT with the polymer [29]. The effects of an aromatic group on the main chain and

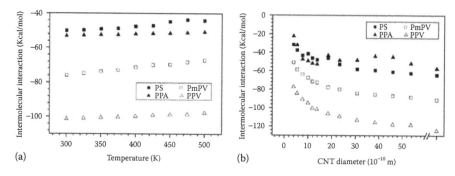

FIGURE 8.3

Influence of (a) aromatic group position in the polymeric chain and temperature and (b) SWCNT diameter on the extent of polymer/CNT interaction. (From Yang, M. et al., *J. Phys. Chem. B*, 109, 10009–10014, 2005.)

environmental temperature on the CNT/polymer interaction energy have been shown in Figure 8.3a. Improved intermolecular interaction is achieved with polymers having an aromatic group in their backbone, such as poly (m-phenylenevinylene-co-2,5-dioctyloxy-p-phenylenevinylene) (PmPV) and poly(p-phenylenevinylene) than polymers with aromatic side groups like poly(phenylacetylene) and polystyrene. Environmental temperature has little effect on the extent of CNT/polymer interaction, as can be seen from Figure 8.3a. However, the CNT diameter has a strong influence on the interfacial interaction, as can be seen from Figure 8.3b.

The characteristics of the bulk polymer may be significantly different from the polymer at the interphase. Enhanced interfacial bonding of the CNT with the surrounding polymer results in the development of a polymer coating around the surface of the nanotube. Improved adhesion of the polymer with the nanofiller leads to formation of a thicker coating. Formation of a PmPV coating of 25 nm thick on two CNTs can be seen from Figure 8.4 [30]. In addition, polymer dendrites have also been noticed at the nanotube end. When a conjugated polymer interacts with a CNT, polymer coating on the surface of CNT, dendrite formation at the defect sites, and polymer wrapping around the nanotube are commonly observed phenomena [31,32]. Due to the modified structure of the polymer in the interphase region, it exhibits significantly higher strength than that of the bulk polymer. As a consequence, a higher energy is required for the CNT pull-out event [33], which indicates a higher fracture toughness of the material.

8.2.3 Nanofiller/Polymer Interface Engineering

The extent of stress transfer from the polymer to the reinforcement is the decisive factor for the polymer matrix composites. In this regard, most of the properties of a composite material are governed from the type and health of the interface. Owing to the enormous specific surface of a nanofiller, the

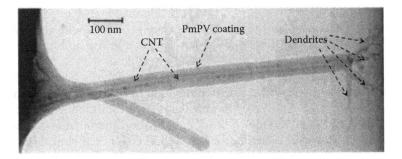

FIGURE 8.4
TEM image of two MWCNTs coated with PmPV. The dark lines represent the MWCNT and at
the end, dendritic growth of the polymer can also be seen. (From McCarthy, B. et al., *J. Phys.
Chem. B*, 106, 2210–2216, 2002.)

potential expected benefit lies in having a huge interfacial area per unit volume of the material for stress transfer. This implies a better reinforcement efficiency with a minute quantity of the reinforcing nanofiller. In addition to this size of interfacial area, the strength of the interface is also equally important. A stronger interface facilitates stress transfer to a better extent by restricting interfacial debonding. Hence, to avail the potential strength of the nanofiller, strong interfacial bonding is required, which can be achieved by effective physical and/or chemical tailoring of the CNT/polymer interface. Due to the hexagonal graphene structure of the carbon atoms in the CNT, most of these carbon atoms are thermodynamically stable. Thus, they have less affinity to interact with the foreign molecules. Hence, when mixed with polymer, only Van der Waals kind of weak physical interaction takes place between a CNT and a polymer. To strengthen the CNT/polymer bonding, various techniques have been evolved to trigger an improved interaction. "Surface functionalization" and "Surface modification" are some of the pronounced terms used in the case of a CNT to modify its surface characteristics. Functionalization essentially means binding the polymer with the CNT by a functional bridge. Functionalization has been classified into two groups based on the type of interaction of the nanotube with the functional group: (1) chemical or covalent functionalization indicates the presence of chemical bonding between a CNT and a polymer, whereas (2) physical or non-covalent functionalization is based on the physical π–π kind of interaction between them [34].

8.2.3.1 Chemical Functionalization

Chemical functionalization is basically used to have a chemical linkage between the CNT and polymer. This is done broadly by establishing chemical bonding between the carbon atoms on the surface of a CNT and one end of the active functional group. The other end of the functional group remains chemically active. When this functionalized CNT is added to the polymer, the active site of the functional group is bonded with the polymer [35–37] and, hence, the functional group acts as a bridge to bind the CNT with the polymer [38].

Various theoretical [39] and experimental [40] analyses have proven the beneficial impact of chemical functionalization of CNT on the performance of the CNT-embedded polymer composites. Chemical functionalization has further been classified as sidewall functionalization and open-end function-alization [41]. However, there exists a profound effect of chemical function-alization of a CNT on CNT/polymer interfacial strength. Even with a very minute fraction of chemically functionalized carbon atoms (~1%), chemical functionalization is capable of increasing the interfacial shear strength by an order of magnitude [42]. In addition, this also leads to a substantial decre-ment in the critical length of the nanotube required for effective stress trans-fer from the polymer matrix to the nanotube. Apart from these aforesaid advantages, surface functionalization also helps in achieving a better extent of dispersion of the nanotubes in the polymer matrix [41].

The process of chemical functionalization of a CNT changes the hybrid-ization state of carbon atoms from a sp^2 to a sp^3 state, enabling the carbon atoms to have bonding with the functional group. One of the potential ways to have a chemically bonded CNT/polymer is *in-situ* polymerization. During this process, the monomers, having free radicals, interact with each other and with the functional groups attached to the CNT as well [43,44]. For nonreactive polymers, the ozone-mediated technique is one of the ways for functionalization [45]. During the ozonization process, the nonreactive polymers undergo chemical reactions, yielding hydroperoxide or alkylper-oxide groups. These groups, under the action of heat, convert to free radicals, which further act as suitable links to have chemical bonding of the polymer with the carbon atoms in the CNT. Other suitable techniques have also been evolved to have an improved CNT/nonreactive polymer, such as microwave irradiation [36,46], plasma-enhanced chemical vapor deposition, and so on.

Chemical functionalization of CNTs with various active functional groups has been an active area of research. A possible technique of chemical func-tionalization of a CNT and its effect in a CNT/polymer composite has been represented in Figure 8.5 [13]. As mentioned earlier, the process of chemical functionalization of a CNT starts with making some of the surface carbon atoms of the nanotube active, which can further be bonded to the functional group. Oxidation is a technique to break the double bond between adjacent carbon atoms of the CNT and makes them active. In addition, it also attaches the active carbon atoms with oxygen containing functional groups, which generally come from the oxidative medium. Acidic solutions are commonly used oxidizing agents. A few examples of acidic media are a mixture of hydro-chloric acid and sulphuric acid [47], a mixture of nitric acid and sulphuric acid [37,48,49] and nitric acid [50] and so on. Due to this process, oxygen contain-ing functional groups like hydroxyl, carbonyl, carboxyl, or ester are attached

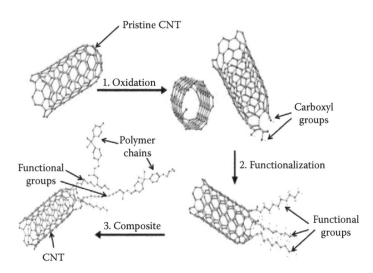

Pristine CNT

1. Oxidation

Carboxyl groups

Polymer chains

Functional groups

2. Functionalization

Functional groups

3. Composite

CNT

FIGURE 8.5

Schematic of CNT functionalization. (From Fiedler, B. et al., *Compos. Sci. Technol.*, 66, 3115–3125, 2006.)

to the active carbon atoms of the nanotube [34]. A very commonly used oxidizing agent is the mixture of concentrated nitric acid and sulfuric acid in a 1:3 ratio. Ramanathan et al. [48] have oxidized CNTs using this acid mixture by sonicating them for 3 h at 40°C. After its pH neutralization, the CNTs were filtered to obtain oxidized CNT. To confirm the oxidation of the CNTs, Fourier transform infrared spectroscopy (FTIR) has been used. In the case of unmodified/pristine CNT, the weak band at ~3,400 cm^{-1} (Figure 8.6a) represents an O–H bond in the CNT, which might be due to partial atmospheric oxidation or moisture content in the sample. However, this peak has been sharpened significantly in the case of oxidized/carboxylated CNTs, as can be seen from Figure 8.6b. In addition, a new peak has been evolved at ~1,720 cm^{-1}, indicating the stretching of the carbonyl (C=O) bond of the carboxyl (COOH) group present on the surface of the CNT. This carboxyl group has been attached to the CNT from the oxidizing agent. In addition, X-ray photoelectron spectroscopy results also indicated the increment in oxygen concentration from 3.4 at.% to 8.75 at.% due to this oxidation treatment. This oxidized/carboxyl functionalized CNT can be used for fabricating a polymer composite with chemical bonding between a CNT and a polymer. Otherwise, the oxidized CNT can be further treated with other chemical agents to develop other functional groups on the CNT surface having a stronger affinity for polymer. Here, a process has been narrated for amine functionalization of CNTs (Figure 8.5). For amine functionalization, Ramanathan et al. [48] have sonicated the oxidized CNTs with an amine source, that is, ethylenediamine and a coupling agent, that is, N-[(dimethylamino)-1H-1,2,3-triazolo[4,5,6]pyridin-1-ylmethylene]-N-methylmethanaminium hexafluorophosphate N-oxide for 4 h.

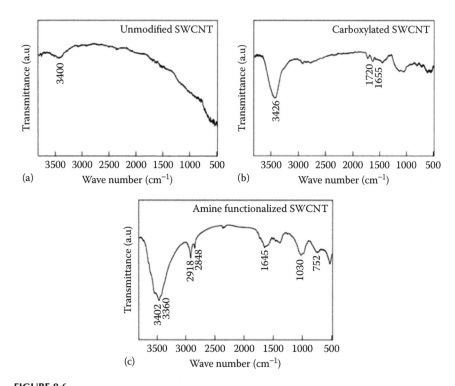

FIGURE 8.6
FTIR spectra of SWCNT in (a) unmodified, (b) carboxylated, and (c) amine-modified conditions. (From Ramanathan, T. et al., *Chem. Mater.*, 17, 1290–1295, 2005.)

The resulting amino-functionalized CNTs were also characterized by FTIR (Figure 8.6c). The fingerprint of the presence of an amine group was detected by the NH_2 stretch peak obtained at 3,402 cm^{-1}. In addition, a weak peak at 3,360 cm^{-1} revealed the symmetric stretching of the amine group. The in-plane scissoring kind of bending mode corresponding to the primary amine group has a broader shape than all other peaks in the near vicinity. However, the wide peak at 752 cm^{-1} indicates the out-of-plane bending of the primary amine. In addition to these bonds, presence of a C–N bond was also confirmed from the peak at 1,030 cm^{-1}. Additional peaks at 2,918 cm^{-1} and 2,848 cm^{-1} correspond to the stretching of the CH_2. When such a type of functionalized CNTs are added to the polymer, the free end of the functional group gets attached to the polymer chemically, which thus integrates the polymer with the CNTs (Figure 8.5).

The beneficial impact of amine functionalization on the CNT/polymer composites has been reported by several authors [51–53]. As reported by Gojny et al. [54], an amino functionalized CNT-based polymer composite exhibits better tensile strength, modulus, and failure strain than its counterpart pristine one. As already discussed, improved interfacial adhesion

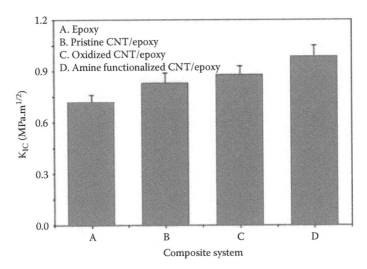

FIGURE 8.7
Fracture toughness of neat epoxy and epoxy nanocomposites. (From Jin, F.-L. et al., *Mater. Sci. Eng. A*, 528, 8517–8522, 2011.)

between a CNT and a polymer due to functionalization helps in achieving better mechanical performance. In addition, a better extent of dispersion due to CNT functionalization is also another possible reason for improved mechanical behavior. Jin et al. [55] have also shown the effect of pristine, oxidized, and amine functionalized CNT on the fracture toughness of CNT/ epoxy nanocomposites, as shown in Figure 8.7.

However, the functional groups are not restricted to only hydroxyl and amine groups. CNT functionalization has also been reported by several other functional groups, like epoxide [37,56], ester [57], silane [58,59], and so on. Kathi et al. [60] have reported the beneficial impact of the silane functionalization of a CNT on the flexural response of the polymer composite, as shown in Figure 8.8. A silanized CNT/epoxy composite exhibited better flexural performance than both an oxidized CNT/epoxy composite and a neat epoxy. Most of these functional groups are polar in nature, and, hence, the extent of agglomeration due to Van der Waal's forces is restricted between adjacent CNTs, resulting in a better dispersion of the CNTs in the polymer composites. In addition, the functionalized CNTs are hydrophilic in nature, compared to their counterpart pristine CNTs. Hence, the wettability of the functionalized CNT by the polymer becomes better and thus a better interfacial interaction is expected.

Several literature have also reported the effect of functionalized CNT on the mechanical performance of CNT-embedded FRP composites. Kim et al. [61] have studied the effect of silanized CNT and oxidized CNT reinforcement in carbon fiber/epoxy composites. Use of 2% silanized CNT in a carbon fiber/ epoxy composite yielded 15% and 55% more improved flexural strength than a control carbon fiber/epoxy and 2% oxidized CNT/carbon fiber/epoxy

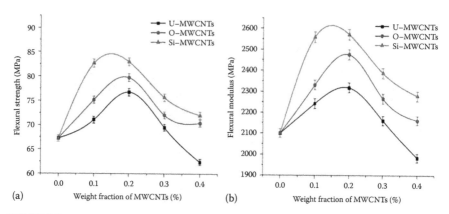

FIGURE 8.8
Effect of pristine, oxidized, and silanized CNT on (a) flexural strength and (b) flexural modulus of epoxy. (U–MWCNT is the unmodified MWCNT, O–MWCNT is the oxidized MWCNT, and Si–MWCNT is the silane functionalized MWCNT). (From Kathi, J. et al., *Compos. Part A Appl. Sci. Manuf.*, 40, 800–809, 2009.)

composite, respectively. A flexural modulus of a silanized carbon fiber/epoxy composite was found to be higher than the other two composites.

8.2.3.2 Physical Functionalization

Chemical functionalization of CNTs has been proven to be a successful evolution as far as the interfacial interaction between a CNT and a polymer is concerned. However, there also exists certain drawbacks associated with this process. During the sonication and other high intensity energy mixing processes for the chemical functionalization of nanotubes, several defects are formed on the sidewall and ends of the nanotubes. Considering the extremity of chemical functionalization, one may end up with disintegration and/or length shortening of the nanotubes, losing their potential performance. Use of very strong acid or oxidizing agents also deteriorates the structure of the nanotubes. In applications where the electronic structure of nanotubes has a vital role to play, physical functionalization is a more viable option than a chemical one. In the physically functionalized state, the sp^2 conjugated state of the carbon atoms remains preserved.

For dispersing nanotubes in a polar polymer matrix, use of an amphiphilic surfactant is an efficient technique. The amphiphilic surfactant has two parts: (1) hydrophilic and (2) hydrophobic. The hydrophilic part of the surfactant is attached to the polar group of the polymer. The hydrophobic part then has a tendency to interact with the nanotube [62]. However, efficiency of this interfacial interaction and dispersion is governed by the type of the amphiphilic surfactant, that is, hydrophilic and hydrophobic groups. A surfactant, Polyoxyethylene 8 lauryl, has been reported to exhibit a positive impact on the extent of CNT dispersion and CNT/epoxy interfacial adhesion in the

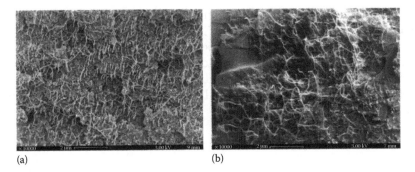

(a)   (b)

FIGURE 8.9
Fracture morphology of CNT/epoxy composite (a) with surfactant and (b) without surfactant. (From Gong, X. et al., *Chem. Mater.*, 12, 1049–1052, 2000.)

CNT/epoxy composite [63]. The scanning electron microscope (SEM) images in Figure 8.9 indicate the wavy entangled CNTs in the case of an unmodified CNT/polymer composite, whereas surfactant modified CNTs show a better extent of dispersion. The glass transition temperature (T_g obtained from dynamic mechanical analysis) of the epoxy was improved by 9°C due to 1% CNT embedment, whereas with the same content of a surfactant-treated CNT exhibited 25°C higher T_g than neat epoxy.

A compatibilizer named P2 has been developed by Mandal et al. [64] via atom transfer radical polymerization containing thiophene moiety and poly(dimethylamino ethyl methacrylate). A suspension with more than 3 h of stability was observed in the case of P2-CNT in dimethylformamide. With 0.05% CNT in poly(vinylidene fluoride) (PVDF), the P2-treated CNT/PVDF composite showed 2–3 times superior tensile modulus, strength, and toughness than untreated CNT/PVDF composite.

In the case of a conjugated polymer, wrapping of the polymer layer around the CNT surface due to π–π interaction is an improved technique of non-covalent functionalization. Due to the polymer layer on the CNT surface, the Van der Waals interaction between adjacent CNTs is disrupted, resulting in a reduced extent of agglomeration [65].

8.2.4 Degree of Dispersion

The degree of dispersion, which possesses a strong influence on the properties of the nanocomposite, is to be quantified for the sake of a better comparison. A better light to the difference in concepts of "macrodispersion" and "nanodispersion" of CNTs was evolved from the NASA- and National Institute of Standards and Technology-sponsored workshop to address the issues of CNT purity and dispersion [66]. Macrodispersion is defined as the dispersion of the small CNT bundles and nanodispersion is defined as the dispersion of the individual CNTs by the splitting of the bundles.

For several critical applications including biosensors, a nanoelectronic devices dispersion of individual CNTs is very important [66]. A dimensionless quantity, X may be used to express the degree of dispersion. Various expressions have been formulated to define X. X_1 may be referred to as the total fraction of CNTs that is in an individually dispersed condition among the total CNTs added or the probability that the entity will have a single CNT ($p(1)$):

$$X_1 = p(1) \tag{8.11}$$

Another way of expressing the degree of dispersion (X_2) is considering the average number of CNTs in a bundle:

$$X_2 = \frac{1}{\text{avg}(z)} \tag{8.12}$$

where avg(z) represents the average number of CNTs per bundle. This average is often determined from the root mean square of the bundle diameter.

The combination of the above-mentioned X_1 and X_2 represents another measure of dispersion of CNTs (X_3):

$$X_3 = \frac{p(1)}{\text{avg}(z)} \tag{8.13}$$

For a perfect nanodispersion, all of these measures of dispersion, that is, X_1, X_2, and X_3 become unified. However, appropriate experimental technologies should be developed for a proper quantification of all of these degrees of nanodispersion.

8.3 Fabrication of Polymer Nanocomposites with Carbon Nanotubes

8.3.1 Thermoplastic Polymer-Based Nanocomposites

Several methods have been designed for fabricating thermoplastic-based nanocomposites. The proper mixing and processing temperature are the major criteria for such system. The various techniques available for fabricating such polymer nanocomposites are described below.

8.3.1.1 Melt Processing of Nanocomposites

This technique of nanocomposite synthesis is mostly involved with thermoplastic polymer, which can be solidified from its liquid state.

Certainly processing takes place when the polymer exists in a viscous liquid form. Hence, the operating temperature for amorphous polymers usually is more than their glass transition temperature, while for semi-crystalline polymers, it is more than the melting temperature, where the polymer pellets are in viscous liquid form. Once sufficient fluidity is achieved in the polymer, the required amount of the CNT can then be mixed. Subsequently, injection molding, compression molding, or extrusion can process different sized and shaped components. In addition to this, care must be taken to optimize the processing parameters, as nanofillers have a tendency to modify the flow behavior of polymers. The major advantage associated with this process is that it can be used for a mass scale production.

Jin et al. [67] have reported the fabrication of a poly(methyl methacrylate) composite reinforced with multiwalled carbon nanotubes (MWCNT). A predetermined weight of MWCNT and PMMA were mixed in a laboratory mixing molder at a speed of 120 rpm, at 200°C for 20 min. The mixture was then converted to films by applying a pressure of 8–9 MPa at 210°C for 5 min. Good dispersion and absence of any agglomeration was confirmed by transmission electron microscope (TEM) images.

Some advanced melt processing techniques are discussed below.

8.3.1.2 Injection Molding

The simplest method of nanofiller dispersion in a polymer is probably through a MiniMax injection molder (Custom Scientific Instruments, Inc.), where the thermoplastic polymer pellets and nanofiller are added and mixing takes place by the rotation of the rotor.

The schematic of an injection-molding setup is shown in Figure 8.10. Depending on the melting temperature of the polymer, the temperature of

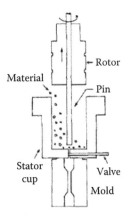

FIGURE 8.10
Schematic of a MiniMax injection molder. (From Koo, K.-K. et al., *Polym. Eng. Sci.*, 25, 741–746, 1985.)

the stator cup is set. The amount of mixing time must be optimized to prevent CNT breakage caused from long-time mixing. The effectiveness of this process can be enhanced by several techniques, like the addition of steel balls [69] and Teflon disks [70] into the mixer. During processing, several times the rotor pin is moved up and down to facilitate radial mixing. After completion of this process, the polymer/nanofiller suspension is then injected into a mold through the valve by moving the rotor pin down, as shown in Figure 8.10.

8.3.1.3 Single-Screw Melt Extrusion

Single-screw melt extrusion [71,72] is another convenient process for dispersing nanofillers in a thermoplastic polymer. The screw must have an optimum length-to-diameter ratio to accommodate the feed and to supply polymer melt at its correct state when it enters to the die. Mixing in a single-screw extruder takes place in several steps: (1) flow of the polymer pellets and CNT from the hopper to the screw channel, (2) compaction of the charge, (3) melting of the polymer pellets, (4) homogeneous mixing, and (5) supply of the melt polymer/CNT suspension to the die. Further, the single-screw extruder has been divided into various zones: (1) a feeder zone, (2) a compression zone, and (3) a metering zone, as shown in Figure 8.11. In the feeder zone, the polymer pellets and CNT are added through a hopper. The temperature of the barrel is set as per the melting temperature of the polymer and the required viscosity. The charge materials are transported through

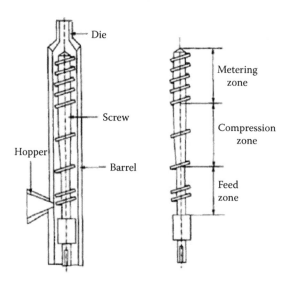

FIGURE 8.11
Schematic representation of single-screw extruder and its various zones. (From Kim, S.J. and Kwon, T.H., *Adv. Polym. Technol.*, 15, 41–54, 1996.)

the rotating screw and, because of a high temperature, the polymer starts melting. In the compression zone, the volatile matters are directed toward the hopper and then escaped from the system. In the metering zone, mixing takes place to achieve uniform composition/dispersion and temperature.

The twin-screw extruder [73–75] is another commonly used laboratory technique for preparation of thermoplastic polymer nanocomposites. Instead of using a single screw, two intermeshing and corotating screws are used for enhanced transportation, compression, shear, heating, pumping, and so on. Superior mixing takes place due to the intermeshing screws.

8.3.1.4 Solution Processing of Nanocomposites

Both polymer and nanofiller are mixed with a suitable solvent through a mechanical mixing technique. Mechanical mixing imparts sufficient energy to the system for deagglomeration of the nanofillers and effective dispersion in the solvent. Sonication, shear mixing, magnetic stirring, or a combination of these mixing techniques is usually used for obtaining a uniform dispersion. The amount of mixing time plays an important role in this context. After getting the desired mixing in the system, the controlled removal of the solvent yields dispersed nanofillers in the polymer matrix.

Several pieces of literature are available showing the processing of nanofiller-reinforced composites through this method. Jin et al. [77] have reported the fabrication of an arc discharged carbon nanotubes-reinforced thermoplastic polymer (polyhydroxyaminoether). First, the MWCNT was sonicated in chloroform (solvent) at room temperature for 1 h. Then, polyhydroxyaminoether was added to this suspension and further sonication was carried out for an additional hour. The suspension was then poured into a Teflon mold and air dried, which yielded the composite film.

Shaffer et al. [78] have reported the successful fabrication of CNT/poly(vinyl alcohol) composites through a solution processing route. Catalytically grown nanotubes were mixed with an aqueous poly(vinyl alcohol) solution and, later, water is evaporated to obtain the nanocomposite.

Several authors [79,80] have reported the fabrication of a MWCNT/polystyrene composite by a high-energy sonication method followed by solution evaporation. Initially, polystyrene was mixed with toluene. The MWCNT was dispersed in toluene separately by high-energy sonication, which was then mixed with a polystyrene–toluene solution. Bath sonication was carried out for further dispersion to take place. Then the mixture was transferred to a mold where the toluene was allowed to evaporate to obtain an MWCNT/polystyrene composite.

Ruan et al. [81] reported the fabrication of an MWCNT/polyethylene using xylene as the solvent. MWCNT was first dispersed in xylene using a magnetic stirrer for 2 h, followed by 2 h of sonication at an ambient temperature. A separate mixture of polyethylene and xylene was made and then mixed with a MWCNT–xylene dispersion at 140°C.

8.3.2 Thermosetting Polymer-Based Nanocomposites

Most of the structural polymeric composites are based on thermosetting polymers. The widely used polymeric composites for load carrying applications are based on epoxy resin. The starting epoxy resin usually exists in liquid form that, upon the addition of a curing agent (hardener), solidifies through the formation of three-dimensional cross-linking bonds. The most important feature of a solid thermosetting polymer is its irreversible phase transformation, that is, upon heating it does not melt, rather, it decomposes above a certain temperature. Hence, all of the processing steps for the synthesis of a polymer composite are to be completed before solidification occurs. For making a nanofiller-reinforced polymer composite, usually the nanofiller should be dispersed properly in the polymer prior to the addition of a hardener. Once the hardener is added, curing takes place and the material solidifies.

Several techniques, alone or combined, are used for getting uniform dispersion of a nanofiller in thermosetting resins. Some of these techniques are discussed in the following sections.

8.3.2.1 Ultrasonic Mixing

This technique may be used to disperse CNTs in liquid polymers (thermosetting or thermoplastics). High local impact energy is imparted by ultrasonic equipment, which introduces lower shear energy into the system. A simple solution–evaporation method has also been applied to prepare homogeneously dispersed MWCNTs in a polystyrene matrix using a high-energy sonication technique without sacrificing the integrity of the CNTs [79]. Another beneficial effect, which can be drawn using a sonicator, is deagglomeration of the agglomerated CNTs prior to adding them into the polymer. Initially, the CNTs are dispersed in a solvent like acetone, and this suspension is then sonicated, where the imparted vibrational energy makes the CNTs move away from each other. The extent of deagglomeration, again, depends on the type of CNTs. The higher the specific surface area, the higher is the tendency for agglomeration. Hence, in the case of single-walled CNTs (SWCNTs), it's very difficult to get a uniform dispersion. Further, as the imparted energy is very high, it may eventually damage the nanofiller shape and size. Defects like buckling, bending, and dislocations may become very significant due to ultrasound treatment of CNTs [82]. After an optimized period of sonication, the suspension is then mixed with the polymer and just heating evaporates the solvent.

Song et al. [83] were able to fabricate an MWCNT/epoxy composite, where dispersion was achieved by sonication. The MWCNT aggregates were sonicated in ethanol for 2 h. The CNT/ethanol mixture was then mixed with epoxy, which was further sonicated for 1 h at 80°C and kept under vacuum for 5 days for removal of air bubbles and ethanol. Then, the required amount of hardener was added and mixed well with the help of magnetic stirring under sonication for 15 min. The use of ethanol as a solvent resulted in a nice dispersion of the nanofiller in the epoxy matrix and subsequently exhibited superior mechanical properties.

8.3.2.2 Mechanical Mixing

One of the common conventional techniques to disperse nanofiller in ther-mosetting resin is through mechanical stirring. Predetermined quantities of resin and nanofiller are taken and mixing is carried out through mechanical agitation. In this technique, the process variables include propeller size, tem-perature, mixing speed, and time. A better dispersion can be obtained with a low viscous resin and a smaller volume of the mixture.

Sandler et al. [84] prepared a CNT-reinforced epoxy composite, where ini-tially deagglomeration of CNTs was done by sonication, followed by mix-ing with the epoxy resin by stirring. The stirring was continued for 1 h at 2,000 rpm and at a temperature of 80°C. A higher temperature lowers the resin viscosity and, hence, facilitates mixing. Experimental observations refer a better scale of dispersion in the case of the MWCNTs than the SWCNTs. But, again, excessive stirring may result in agglomeration or flocculation, due to elastic interlocking mechanisms and frictional contacts [85]. In the case of thinner nanotubes, a larger shear force is required for effective dispersion because of the high specific surface area.

Allaoui et al. [86] have reported the fabrication of an MWCNT/epoxy composite by a two-step process. To reduce the extent of agglomeration/ entanglement in an MWCNT, first it was dispersed in a methanol bath and stirred through magnetic agitation. Then, the methanol was evaporated completely and the MWCNT thus obtained was mixed directly to the epoxy/ hardener mixture, which is then homogenized manually. Then, the mixture was kept between two metal plates at a certain pressure to minimize poros-ity due to curing.

8.3.2.3 Calendering

The three-roll calendering mill is one of the modern and promising approaches for getting uniform dispersion of nanofillers in a polymer resin. Gojny et al. [54] have reported a CNT/epoxy nanocomposite through this technique. The gap between the rolls was kept at 5 µm. The speed of rota-tion of the rolls from the starting to the exit point was set in an increas-ing order (entry, middle, and exit rolls at 20, 60, and 180 rpm, respectively). The manually mixed CNT and the polymer suspension were then fed to the mill. A huge shear force exhibited by the suspension during its 2 min stay in the mill was sufficient for having a nice dispersion of CNTs in the polymer matrix. Then, to the dispersion, a required amount of hardener was added for curing.

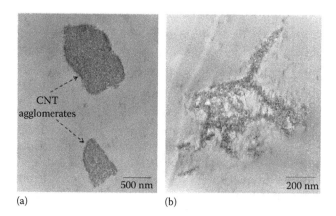

(a)   (b)

FIGURE 8.12

TEM micrograph of (a) condensed DWCNT agglomerate in epoxy even after sonication and (b) exfoliated open structure of DWCNT in epoxy after Calendering. (From Gojny, F.H. et al., *Compos. Sci. Technol.*, 64, 2363–2371, 2004.)

Gojny et al. [54] have also reported the comparative dispersion of amino functionalized double-walled CNT (DWCNT) in an epoxy matrix by sonication and calendering routes. Calendering gives a better degree of dispersion, as shear force imposed by the rolls is uniform throughout the resin, whereas in the case of sonication, local shearing takes place. In calendering, the exfoliation of CNT takes place, which in turn opens up the condensed CNT structure into an open structure, as shown in Figure 8.12, which thus improves the degree of dispersion.

Another viable solution to achieve superior dispersion in the case of a CNT/polymer (thermoplastic and thermosetting both) composite is by using an appropriate compatibilizer. The addition of maleic anhydride grafted styrene-(ethylene-co-butylene)-styrene copolymer compatibilizer in an ultrahigh molecular mass polyethylene/CNT composite has been reported to be beneficial for the uniform dispersion of CNTs [87]. The coating of PVDF on the MWCNTs produced by sonicating MWCNTs in PVDF (acts as compatibilizer) resulted in a superior dispersion in the poly(methyl methacrylate) matrix processed by melt-blending [88]. Surfactants have also been reported to act as a dispersing agent. Polyoxyethylene-8-lauryl, a nonionic surfactant in a CNT/epoxy composite, resulted in a higher T_g (from 72°C to 88°C) value due to better dispersion and interfacial adhesion [63].

The processing technique and parameters strongly influence the microstructural evolution in the resulted composite and, hence, govern the final properties of the composites. Effective dispersion is a stringent criterion to be achieved, but not at the cost of damaging the structure of a CNT. The selection of a proper processing technique depends on several factors, such as the viscosity of polymer, type and content of CNT, and required degree of dispersion. Ultrasonication and stirring are simple and low cost techniques, but may not result in the required state of dispersion. Sonication for a longer duration even may result in damage to the CNTs, reducing their reinforcement efficiency. Calendering results in a better degree of dispersion, but is a relatively high cost instrument and requires the proper skills to better position the rolls, to control their speed, and to clean for repeated use [89]. Another important aspect of calendering is that it may eventually lead to the alignment of the CNTs in the polymer matrix.

Polymer nanocomposites may be fabricated using one or a combination of the above-mentioned techniques to optimize the degree of dispersion.

8.4 Fabrication of Carbon Nanotube-Embedded Fiber-Reinforced Polymer Composites

A nanofiller-reinforced polymer composite may be fabricated using any of the above techniques depending on the polymer and nanofiller characteristics. For making nanofiller-modified FRP composites, the nanofiller may be dispersed in the polymer phase initially as described in the above processes and, further, this nanofiller/polymer suspension may be used as the matrix phase to process a fiber (glass, carbon, Kevlar, alumina, silica, etc.)-reinforced polymer composite by hand lay-up, vacuum bagging, resin transfer molding, and so on. Depending on the type of polymer, the processing technique and conditions (temperature, gel time, etc.) should be decided. However, due to the addition of nanofiller in polymer resin, the curing behavior also gets affected for thermosetting polymer, which is to be kept in mind during fabrication.

For better enhancing the fiber/polymer interfacial properties, CNT may be either grafted onto the fiber surface (by chemical vapor deposition or electrophoretic deposition) or using CNT in the fiber sizing, which has been described later in Section 9.3.

8.4.1 Thermoplastic Polymer-Based Nanophased Fiber-Reinforced Polymer Composites

The convenient and simplest approach is to disperse the CNTs in the polymer resin/melt. The effectiveness of this technique lies in the dispensability of the CNTs in the liquid polymer. Hence, this again requires the polymers melt with sufficient fluidity. As mentioned in Section 8.3, several well-established techniques, for example melt processing and solution processing can be utilized to optimize the dispersion of CNTs in the polymer. This CNT/polymer composite is further utilized as a matrix for the fabrication of FRP composites. The CNT/polymer nanocomposite may be in the form of solid sheets, which can be placed in between the fiber sheets (the fiber might be unidirectional or bidirectional). This assembly of fiber sheets and CNT/ polymer sheets can be placed in the hot press for its further consolidation and integration under the optimized pressure and temperature. Shen et al. [90] used this technique for fabricating a glass fiber/PA6/MWCNT laminate. Initially, thin films (100 μm) of PA6/MWCNT were prepared. There were 10 glass fiber layers and 11 PA6/MWCNT films that were alternatively stacked, heated at 240°C for 10 min, followed by hot pressing for 10–20 min at 0.8–1 MPa pressure. The final laminate contained 49 ± 2 wt% fibers with less than 2% void content. Díez-Pascual et al. [91–93] have also reported the fabrication of a glass fiber/polyetheretherketone (PEEK)/SWCNT laminate using this technique. The melt blending of SWCNT and PEEK was carried out in an extruder operated at a 380 ± 5°C temperature with a rotor speed of 150 rpm for 20 min [94]. These small extrudates were further compressed in a hot press to prepare 0.5 mm thick films at a certain pressure. A stack was then prepared by putting alternate layers of glass fiber and SWCNT/PEEK films. The consolidation of this stack was then carried out in a hot press at a 380 ± 5°C temperature at three increasing pressure steps of 10, 40, and 130 bars. The resulting laminates contain 63 ± 1 wt% fibers having a density of 1.82 ± 0.04 g/cm^3.

8.4.2 Thermosetting Polymer-Based Nanophased Fiber-Reinforced Polymer Composites

Various methods have been developed for fabrication of thermoset polymer-based FRP composites, for example hand lay-up, resin transfer molding (RTM), and so on. These techniques can be further extended to fabricate CNT-incorporated FRP composites [95–98]. In most of these techniques, a low viscosity resin is a prerequisite, which enables the impregnation of large fiber preforms for the fabrication of large components. The modification of the polymer resin by the incorporation of any strengthening/toughening agent like rubber particles, thermoplastic particles/fibers [99], and nanoparticles like CNTs can significantly modify the rheology of the resin system, such as viscosity and gel time. Another important concern in all the liquid molding techniques with nanofiller modified resin is the "filtration" of the resin by the fibers. Filtration leads to heterogeneous distribution of the nanofillers in the final composite, promoting distinct interfaces at different locations. Reia da Costa et al. [100] have successfully demonstrated the fabrication of both glass FRP (GFRP) and carbon FRP (CFRP) composites using a CNT-modified epoxy resin. A three-roll calendering mill initially dispersed CNTs in an epoxy resin. This CNT/epoxy suspension was then impregnated into the fiber stack by the RTM process. To reduce the viscosity of the CNT/epoxy mixture, it was heated to a temperature of 40°C, followed by degassing for some time to remove entrapped gases. Then, hardener was added to the suspension, stirred, and followed by degassing for some time.

The resin was then transferred to the preheated piston of the RTM and mold filling was carried out at a 2 bar pressure. Then, the resulted samples were cured.

Most of these techniques can be used for fabricating a CNT-embedded FRP composite with a low CNT content. For introducing more CNTs into a FRP composite, fiber modification is the possible way.

8.5 Mechanical Performance of Carbon Nanotube-Embedded Polymer Composites

The mechanical properties of the polymer nanocomposites not only depend on the type of CNT and polymer, but also depend on the processing parameters and the interfacial interaction. Very often it is noticed that improper dispersion leads to poor mechanical performance and, hence, the potential exploitation of the properties of CNT is not achieved. There exists a vast literature reporting the mechanical property enhancement by CNT incorporation by various techniques.

Allaoui et al. [86] have reported the tensile properties of MWCNT-reinforced epoxy composites. Both Young's modulus and yield strength values were doubled and quadrupled in comparison to pure epoxy, with 1% and 4% MWCNT reinforcement, respectively. One important observation regarding the reinforcement efficiency that was also noticed is the normalized stress (at a certain strain, the ratio of stress exhibited by the composite to the pure epoxy) remains almost constant throughout the entire range of strain, as shown in Figure 8.13.

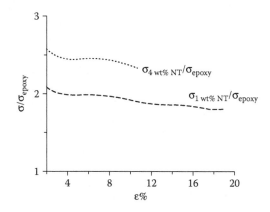

FIGURE 8.13
Normalized stress versus strain for 1% and 4% CNT reinforcement. (From Allaoui, A. et al., *Compos. Sci. Technol.*, 62, 1993–1998, 2002.)

It should be noted that reinforcement efficiency (property improvement per unit wt% of CNT) is quite high with a lower nanotube content, which gradually decreases. The reason for poor reinforcement efficiency at a higher CNT content was attributed to the higher porosity and poor dispersion of CNTs in the epoxy.

Interface helps in transferring the stress from the matrix to the reinforcement. Chemical functionalization of the CNT surface can alter the character and chemistry of the interface in multiscale composite materials. Covalent bonds and/or dipole–dipole interaction can be formed between a polymeric matrix and CNTs by customized functionalization and, hence, this leads to the strengthening of the interface [42,101–102] and also the improvement in wettability and dispersibility of the CNTs.

The addition of CNT to the matrix when studied for interlaminar shear strength (ILSS) enhancement in FRP composites has obtained mixed results in the literature, with the data ranging from no improvement to as high as 27% improvement [96,97,103–104]. Different processing techniques have been developed for fabrication varying from hand layup, autoclave, resin transfer molding, to vacuum-assisted RTM (VARTM). Schulte et al. have reported ILSS improvement of about 20% [96,104] for composites fabricated using RTM, whereas no improvement was noticed in the composite fabricated with VARTM [97]. They concluded that the type of resin used has observable significance on the effect of nanoparticles on the mechanical properties of an epoxy resin. Change in the processing method from RTM to VARTM can also be the reason for the difference in the data obtained. Different distribution of the nanoparticles can be caused by different resin flow patterns.

The tensile and shear strength of nanotube-reinforced composite interfaces were investigated by Hsiao et al. [105] and Meguid and Sun [106] using single shear-lap testing. A significant increase in interfacial shear strength was observed for epoxies with multiwall nanotubes with content between 1% and 5% in comparison to a neat epoxy matrix. Increase in the matrix-dominated ILSS was noted to about 20% by the addition of a small amount of carbon nanotubes (0.3 wt% DWCNT–NH$_2$), while there was no change observed in the tensile properties and still remained fiber dominated [96].

To engineer a structural polymeric composite with CNT incorporation, the possible science of a CNT/polymer interaction must be well understood. Analyzing the interaction and mechanics at the nanoscale still remains a strong challenge for the material scientists. Nevertheless, various possible models and theories of interaction do exist suggesting the various parameters responsible for improvement in strength, toughness, and so on in such composites due to CNT addition. Hence, possible modification in those parameters is the key to achieving superior properties in these composites.

High-resolution electron microscopic observations again have validated some of these models successfully.

8.5.1 Theories and Micromechanisms for Improved Mechanical Performances of Carbon Nanotube-Embedded Polymeric Composites

In addition to strength enhancement, CNT reinforcement has also been proven to be beneficial for improving the toughness of polymeric materials. This has been supported by various theoretical and experimental analyses. The simplest way of visualizing the improved toughness of polymer due to CNT reinforcement is the availability of a significant number of CNTs (due to their small size) ahead of the crack tip, which obstructs the propagation of the crack. However, this mechanism is universal for most of the nanofiller-reinforced polymers. In addition, there are some other toughening mechanisms, which are unique to CNTs due to their geometry. Typical fracture morphologies of neat epoxy and MWCNT-reinforced epoxy nanocomposites have been shown in Figure 8.14. The presence of the riverlines indicate the brittle fracture behavior in the case of a neat epoxy, as shown in Figure 8.14a. However, the fracture surface appearance is quite rough in the case of a nanocomposite, as shown in Figure 8.14b, depicting ductile failure mode [107].

If any event in the material during its deformation is capable of absorbing some energy, then the material is able to withstand a higher load and thus exhibits a higher damage tolerance or fracture toughness. To determine the work done during this pull-out, we can consider a fiber (or CNT) having its free end at a distance of k from the mid-crack plane, which has been pulled out from the matrix by a distance x, as can be seen from Figure 8.15.

30 μm	1000x	30 μm	1000x
(a)		(b)	

FIGURE 8.14
SEM images of fractured surface of (a) neat epoxy and (b) epoxy reinforced with 0.3% amine-functionalized MWCNT. (From Jajam, K.C. et al., *Compos. Part A Appl. Sci. Manuf.*, 59, 57–69, 2014.)

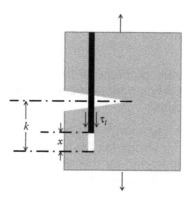

FIGURE 8.15
Matrix crack bridging and pull-out by nanotube.

For doing this, work has to be done against the interfacial shear strength (τ_i). The force at the fiber surface, which is now opposing the pull-out event is $\tau_i\pi d_f (k - x)$ (d_f is the fiber diameter). Work done by this force for a further pull-out distance of dx is $\tau_i\pi d_f (k - x)dx$. Restriction to the crack opening will further take place till the free end of the fiber reaches the mid-crack plane (i.e., pull-out distance becomes k). Total work done by a fiber pull-out (w_{fp}) is given by

$$w_{fp} = \int_0^k \tau_i\pi d_f \left(k - x\right)dx = \frac{\tau_i\pi d_f k^2}{2}$$ (8.14)

This is the work done by a fiber pull-out for a single fiber. For a matrix with N number of fibers (or CNTs) per unit area, the net work done due to a fiber/nanotube pull-out (W_{fp}) is

$$W_{fp} = \int_0^{l_f} w_{fp} \frac{Ndk}{l_f} = \frac{\tau_i\pi d_f l_f^2}{6} N$$ (8.15)

l_f represents the fiber length and the fiber volume fraction $v_f = N\frac{\pi d^2}{4}$
Hence:

$$W_{fp} = \frac{2}{3} \frac{v_f \tau_i l_f^2}{d_f}$$ (8.16)

This expression clearly suggests that the higher the aspect ratio (l_f/d_f), the better is the fracture toughness. As CNTs typically exhibits a very high aspect ratio, this mechanism is in support with the high fracture toughness of the

CNT-reinforced composites. Another important parameter of this above-mentioned expression is the interfacial shear strength of the CNT/polymer interface. Hence, it is also expected that functionalization which enhances the interfacial shear strength is also beneficial for having improved fracture toughness in the composite. This mechanism assumes uniform dispersion of CNTs in the matrix and, hence, its accuracy is strongly dependent on the extent of dispersion of CNTs in the polymer matrix.

The theory of crack pinning can also be validated for enhancement in the fracture toughness of a polymer with strong nanofillers. The speed of crack propagation is lowered due to its interaction with the nanofillers, which act as obstacles for the crack propagation. For passing a series of nanofillers that are ahead of the crack tip, the crack has to bow around the nanofillers through the interface. For a nanofiller-incorporated polymer composite, the fracture toughness (G_{IC}) can be estimated as per the equation mentioned below:

$$G_{IC} = G_{ICm} + \frac{T_L}{s} \tag{8.17}$$

where:
G_{ICm} is the fracture toughness of the neat polymer
T_L is the line tension of the crack tip
s is the interfiller separation

This indicates from Equation 8.4 that for a same volume fraction of filler in the composite, the fracture toughness of the composite would be higher with low diameter filler (hence lower s).

A direct observation of crack bridging by CNT has been shown by Qian et al. [79] by in-situ deformation of a MWCNT/polystyrene nanocomposite film in a TEM by electron beam condensation. A condensed electron beam induced local stress concentration on the film, which was eventually relieved in the form of a crack formation and its subsequent propagation. As per their observation, the probability of a crack formation is higher in the location where the CNT concentration is lower. The CNTs aligned perpendicular to the crack faces were thought to be obstructing the crack propagation rate, as can be seen from Figure 8.16a by a crack bridging mechanism. On the other hand, at a higher crack opening displacement, due to the higher stress concentration on the CNTs which were bridging the crack, eventually fragmented or pulled out from the polymer matrix, as shown in Figure 8.16b.

Gojny et al. [108] have also provided a similar kind of polymer crack bridging by nanotube, as shown in Figure 8.17.

The addition of 0.1 wt% MWCNT in a glass fiber/epoxy composite resulted in 11.5% and 32.7% improvement in the flexural modulus and strength, respectively [109]. Uniform distribution of the CNTs in the matrix

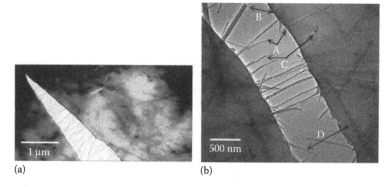

(a)   (b)

FIGURE 8.16
Thermal stress-induced crack nucleation and its further growth as observed inside TEM for an MWCNT/polystyrene composite film showing (a) appearance of craze and (b) MWCNT bridging, pull-out, and breakage. (From Qian, D. et al., *Appl. Phys. Lett.*, 76, 2868–2870, 2000.)

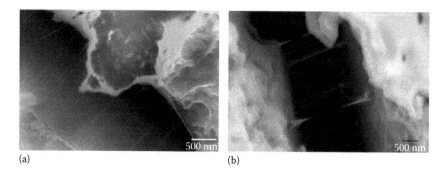

(a)   (b)

FIGURE 8.17
SEM images of the surface crack made by etching DWCNT/epoxy composite, (a) amino functionalized and (b) nonfunctionalized. (From Gojny, F.H. et al., *Compos. Sci. Technol.*, 65, 2300–2313, 2005.)

was stated to be the reason for this improvement, as shown in Figure 8.18a and a′. Above these improvements, a 9.4% higher strain at peak was also obtained in the nanocomposite due to microcrack bridging and CNT pull-out events.

Li et al. [110] reported the effect of a CNT presence at the interphase between the carbon fiber and epoxy. Irrespective of the CNT reinforcement, fracture in these composites was always initiated at the fiber/polymer interface, as shown in Figure 8.19. However, the interface morphology is altered due to the presence of a CNT. Without a CNT, a clean fiber/polymer interfacial debonding could be noticed. On the contrary, the fiber with a CNT was adhered with some polymer, suggesting a better interfacial bonding.

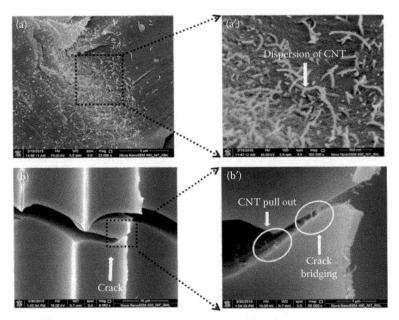

FIGURE 8.18

SEM micrographs of fractured sample of CNT (0.1%)-GE composite tested at room temperature showing (a, a'), dispersion of CNTs in the polymer and (b, b') crack bridging and crack pull-out by CNTs. (From Shukla, M.J. et al., *J. Compos. Mater.*, 0021998315615648, 2015.)

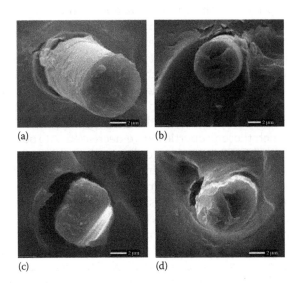

FIGURE 8.19

Interface morphology from a single fiber-composite fragmentation tests (a) raw T700SC, (b) T700SC with COOH–CNTs, (c) raw T300B, and (d) T300B with COOH–CNTs. (From Li, M. et al., *Carbon*, 52, 109–121, 2013.)

8.6 Environmental Sensitivity of Carbon Nanotube-Enhanced Polymer Composites

8.6.1 Temperature

In-service temperature has an influential role in deciding the overall performance of a material. In the case of a polymer matrix composite, the response of the polymer phase and a subsequently generated interphase grossly affect the mechanical performance of the composite at a given temperature. Very often the load bearing capacity of the polymer phase is reduced upon an elevated temperature and a decrement of an order of magnitude may even be noticed once the temperature is raised beyond its glass transition temperature [111–113]. Another important point to consider is the state of the interface at a given temperature. The state of stress at the interface as a result of differential expansion/contraction of the polymer and reinforcement alters the stress transfer efficiency. At low/cryogenic temperature, the polymer gets hardened due to frozen polymeric chains. Various types of nonequilibrium thermal conditioning, such as thermal shock, thermal cycling, and freeze–thaw may alter the performance of the polymer matrix composites [114–116].

8.6.1.1 Cryogenic and Low Temperature Performance

Chen et al. [117] have done a comparative analysis of the reinforcement effect of a CNT in an epoxy at room temperature and liquid nitrogen temperature. In their study, they observed no improvement in tensile strength of an epoxy due to CNT reinforcement at room temperature, as shown in Figure 8.20a. Rather, at a higher CNT content ($\geq 1\%$), the nanocomposite yielded a lower strength than the control sample. On the contrary, a significant enhancement in the tensile strength of an epoxy due to a CNT embedment was observed, when the test was carried out in liquid nitrogen (i.e., 77 K). Both the strength and failure strain were increased with an increase in the CNT content up to 0.5%, after which these properties were decreased. At room temperature, the effectiveness of CNTs to enhance the mechanical performance of the polymer is restricted by the poor interfacial adhesion, which is incapable of transferring load from the weak polymer to the strong CNT. Due to this poor interfacial bonding, CNTs can be easily pulled out of the polymer, losing their potential to reinforce the polymer.

Interestingly, the same CNT/polymer composite yielded a significant improvement in the mechanical performance as compared to a neat polymer, when the test was carried out at 77 K. This behavior can be understood from the thermal stress developed at the CNT/polymer interface at a lower temperature, as shown in Figure 8.21. The coefficients of thermal expansion of the epoxy and CNT are $5.1 \times 10^{-5}\,K^{-1}$ and $0.73–1.49 \times 10^{-5}\,K^{-1}$, respectively [118,119]. This difference in the coefficient of thermal expansion

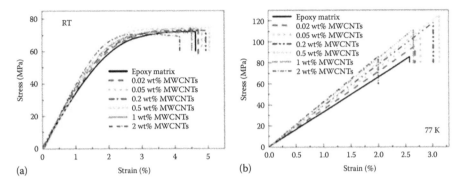

FIGURE 8.20
Tensile testing results for epoxy and MWCNT/epoxy composites with various weight % of MWCNT at (a) room temperature and (b) liquid nitrogen bath (77 K). (From Chen, Z.-K. et al., *Polymer*, 50, 4753–4759, 2009.)

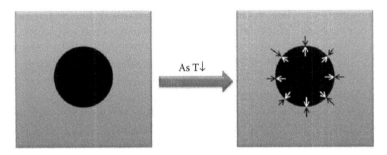

FIGURE 8.21
Effect of lower temperature on the CNT/polymer interface.

of the constituents leads to the development of a thermal stress at the CNT/epoxy interface. The epoxy thus exhibits a compressive or clamping stress on the CNT, which increases the interfacial bond strength [120]. However, all of the materials exhibited brittle failure, due to the cryogenic embrittlement restricting interchain polymer slippage.

Prusty et al. [121] have also investigated the effect of testing temperature on the flexural behavior of glass fiber/epoxy composites filled with 0.3% MWCNT. Flexural properties of the control glass fiber/epoxy and 0.3% MWCNT-filled glass fiber/epoxy composites were compared at different in-service temperatures, as shown in Figure 8.22. The rate of modulus improvement with lowering temperatures was found to be significantly higher for the CNT-filled composite than the control glass fiber/epoxy composite. This can be explained from the same clamping stress point of view as described earlier.

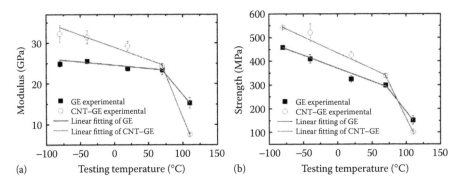

FIGURE 8.22
Effect of testing temperature on flexural (a) modulus and (b) strength of GFRP and CNT–GE (0.3% CNT) composites. (From Prusty, R.K. et al., *Compos. Part B Eng.*, 83, 166–174, 2015.)

8.6.1.2 Elevated Temperature Performance

Rathore et al. [122] have studied the effect of an elevated temperature environment and CNT content on the mechanical performance of the glass fiber/epoxy composite. When tested at an ambient temperature, the glass fiber/epoxy composite with 0.1% MWCNT exhibited the highest strength, which was 32.8% higher than the control glass fiber/epoxy composite. The uniform dispersion of CNTs in the polymer matrix was believed to be the key for this improvement.

However, it can be seen from Figure 8.23 that the strength and modulus of all the nanophased composites decreased at a faster rate than the control glass fiber/epoxy composite as the temperature was increased. This behavior may be explained from the same differential expansion theory. As the CNT-adhered polymer has a tendency to expand faster than the CNT as the temperature is increased, there develops a thermal tensile stress at the

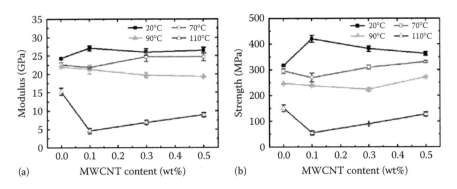

FIGURE 8.23
Effect of CNT content and testing temperature on flexural (a) modulus and (b) strength of CNT–GE composites. (From Rathore, D.K. et al., *Compos. Part A Appl. Sci. Manuf.*, 84, 364–376, 2016.)

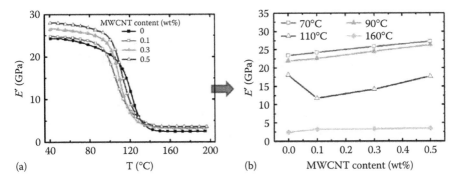

FIGURE 8.24
(a) Storage modulus (*E′*) vs. temperature (T) at various MWCNT content and (b) *E′* vs. MWCNT at various temperatures. (From Rathore, D.K. et al., *Compos. Part A Appl. Sci. Manuf.*, 84, 364–376, 2016.)

CNT-adhered polymer. This thermal tensile stress favors interfacial debonding and thus reduces the reinforcement efficiency of the nanophased composites at elevated temperatures. Interestingly, the nanophased composite, which was having the highest strength at ambient temperatures (i.e., 0.1% MWCNT), was found to be having the lowest strength at 110°C. This behavior has been explained on the basis of a net effective CNT/polymer interfacial area of the nanophased composites. Due to a uniform dispersion, a 0.1% MWCNT-containing glass fiber/epoxy composite has the highest interfacial area. As a result of this, the total interfacial thermal stress per unit volume is at maximum in that composite. Hence, this composite is more prone to interfacial microcrack formation and, consequently, exhibits poor mechanical performance at elevated temperatures. The storage modulus (*E′*) also follows a similar trend as that of flexural modulus, as can be seen from Figure 8.24a and b.

8.6.1.3 Nonequilibrium Thermal Loadings

In practice, there are several applications where the in-service temperature fluctuates. Hence, experiments must be designed in such a way to take this factor into consideration. Nonequilibrium thermal loading refers to rapid heating/cooling of the material or cyclic variation of the temperature. The former is known as thermal shock, whereas the latter is termed as thermal fatigue.

Shukla et al. [109] have reported the effect of cryogenic shock and cryogenic conditioning on the mechanical performance of CNT-embedded glass fiber/epoxy composites. When the materials are suddenly exposed to liquid nitrogen, the constituents get a thermal shock. As a result of this shock, a high magnitude of thermal stress is generated at the CNT/polymer interface. This thermal stress may be released by formation of interfacial microcracks, which eventually reduce the load carrying capacity of the material. Upon application

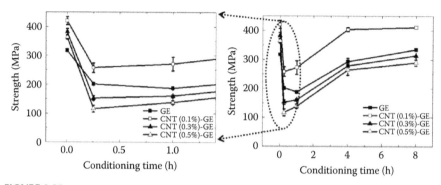

FIGURE 8.25
Variation in flexural strength for glass/epoxy (GE) and CNT embedded GE (CNT–GE) composites with conditioning time in liquid nitrogen. (From Shukla, M.J. et al., *J. Compos. Mater.*, 0021998315615648, 2015.)

of external load, these intermittent microcracks coalesce and propagate, leading to the ultimate failure of the material. Due to this, the extent of strength reduction was found to be higher for nanophased composites than for control glass fiber/epoxy composites, when the samples were tested just after 15 min of cryogenic conditioning, as can be seen from Figure 8.25. However, a prolonged time of cryogenic conditioning leads to matrix hardening and thus results in strength recovery.

Gkikas et al. [123] have evaluated the effect of thermal cycling/fatigue on the thermomechanical properties of neat epoxy and CNT/epoxy composites. A thermal cycle was constituted of 1 h at −30°C and the next 1 h at +30°C. The dynamic mechanical analysis results after 100 such thermal cycles has been shown in Figure 8.26. Before such thermal cycling, the storage modulus of the CNT-embedded epoxy composite was higher than the neat epoxy. However, after 100 thermal cycles, this reinforcement effect was diminished due to the generation of interfacial cracks because

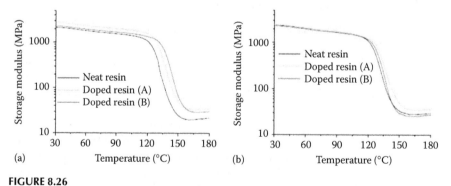

FIGURE 8.26
DMA plots for storage modulus: (a) as manufactured and (b) after 100 thermal shock cycles (a and b represent two different sonication protocols). (From Gkikas, G. et al., *Compos. Part B Eng.*, 70, 206–214, 2015.)

of the repeated differential expansion/contraction of the CNT and epoxy at the interface. The sonication parameter was found to be having very little influence on this behavior.

As several parameters are involved in deciding the effect of nonequilibrium thermal loading on the performance of such nanocomposites, it is quite complex. Due to the presence of different phases, which respond to this temperature fluctuation distinctly, analysis of the effect of such parameters on the bulk mechanical performance of the composite is certainly tedious. However, further investigations must be constructively designed to take all of these factors into account to arrive at a conclusive point.

8.6.2 Hydrothermal and Hygrothermal Exposure

As discussed in Chapter 4, polymers have a tendency to absorb atmospheric moisture. This diffused moisture plasticizes the polymer and, hence, adversely affects its mechanical performance. When a polymer is reinforced with nanoscaled reinforcement, there is a possibility of an alteration in its moisture absorption behavior.

8.6.2.1 Kinetics of Water Ingression

As stated earlier, several modes of moisture ingression in polymer and polymer composites have been discovered and, accordingly, various mathematical models have been proposed [124–130]. Due to the interaction of the water molecules with the polymer chains and interchain bonds, plasticization of a polymer takes place. As a result, the strength and stiffness of the polymer decreases [131]. However, at a later stage of moisture absorption, a recovery in mechanical performance may take place due to possible hydrogen bonding.

Prolongo et al. [131] have reported the influence of CNT reinforcement on the moisture absorption tendency of an epoxy. The trend of water ingression remains unaltered when a CNT is reinforced in an epoxy. However, there is significant decrement in both diffusivity and saturation water content due to CNT reinforcement, as shown in Figure 8.27. This improved resistance against water diffusion may have been from different sources. The presence of hydrophobic CNTs along the water diffusion path hinders the flow of diffusion flux and forces the water molecules to follow a different extended path and, hence, impedes the rate of diffusion. The other factor, which also may contribute toward the water resistance, is the total free volume of the polymer. Free sites in the polymer are the favorable regions for water accumulation. As CNT incorporation reduces the free volume of the polymer [132] (Figure 8.28b), the water absorption propensity of the polymer is thus reduced upon CNT reinforcement.

Starkova et al. [133] have studied moisture conditioning of a CNT-filled epoxy at various ageing temperatures and relative humidity (RH). As per

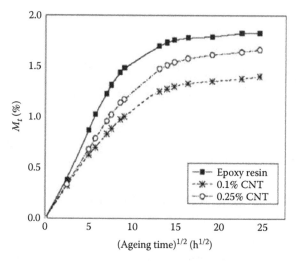

FIGURE 8.27
Water absorption behavior of epoxy and epoxy reinforced with various wt% of amine-functionalized CNTs. (From Prolongo, S.G. et al., *Compos. Part A Appl. Sci. Manuf.*, 43, 2169–2175, 2012.)

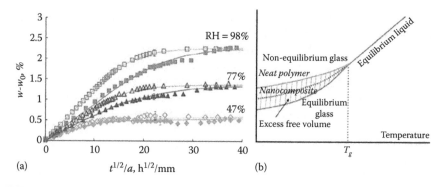

FIGURE 8.28
(a) Water uptake kinetic of neat epoxy (open symbols) and CNT/epoxy composite (closed symbols) at various RH and (b) effect of nanofiller on the free volume of polymer. (From Starkova, O. et al., *Eur. Polym. J.*, 49, 2138–2148, 2013.)

their results, there is a significant decrement in the moisture diffusivity due to the presence of CNTs in an epoxy. However, the saturation concentration remains unchanged, as can be observed from Figure 8.28a. The reduction in free volume of the polymer due to CNT reinforcement (Figure 8.28b) is responsible for the lower diffusivity of a CNT-filled polymer than a neat polymer.

In another study performed by Gkikas et al. [123], the presence of CNTs in an epoxy altered neither the diffusivity nor the saturation content.

Surprisingly, a positive influence of CNT reinforcement has been observed on the water diffusion of a CFRP composite, both in terms of uptake rate and saturation content. The saturation water content has been found to be 1% and 0.6% for the control CFRP and the CNT-filled CFRP composite, respectively. Similarly, the diffusion coefficient of the CFRP was also reduced by 10% due to CNT reinforcement. A higher free volume is expected in the case of the control CFRP composite due to incomplete polymerization, which promotes water ingression. This free volume might have been reduced due to CNT reinforcement in the case of the CNT-filled CFRP composite, thus reducing its water absorption tendency.

8.6.2.2 Mechanical and Thermomechanical Performance after Moisture Ingression

Due to water ingression in a polymer and its composite, various chemical changes take place in the material, which may be either reversible or irreversible in nature. Gkikas et al. [123] have reported the changes in the interlaminar shear strength of CFRP composites with and without CNTs due to water ageing. As discussed earlier, no significant deviation in the water uptake profile could be noticed due to CNT reinforcement in an epoxy. However, the effect of a CNT was quite prominent on the water absorption behavior of a CFRP composite. Even though the CNT-embedded CFRP composite absorbed 40% less water than the control CFRP composite, the difference in ILSS of both of these composites after water conditioning is negligible. The ILSS of both nanophased and control composites was decreased by about 50%–60% due to water ingression.

As discussed in the previous section, Prolongo et al. [131] have reported improved water resistance of an epoxy due to CNT incorporation. Moisture diffusion has significantly altered the thermomechanical response of the composites. As a consequence of plasticization of the polymer, the storage modulus has shown significant decrement, which is evident from Figure 8.29. A two-step fall of the storage modulus can be observed with increasing temperature for both the neat epoxy as well as the CNT/epoxy nanocomposite before and after hygrothermal conditioning. The shallow fall at about 50°C–70°C is termed as β relaxation, which is associated to the movement of the pendant polymeric chains, which may contribute toward further curing of the polymer. As the samples were being conditioned at 55°C for a prolonged time, further post-curing took place, which is evident from the increment in the β relaxation temperature. The deep fall at about 170°C–180°C represents the onset of the glass transition behavior of the polymer. In addition, the tan δ peak gets shortened and broadened, at the same time shouldering also takes place due to hygrothermal conditioning. The absence of chemical linkage between water molecules with the host polymer may be the reason behind this. However, this tendency is lower in the case of a CNT/epoxy nanocomposite, due to absorption of less water.

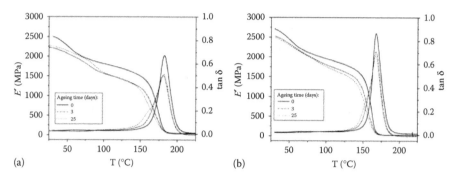

FIGURE 8.29
DMA plots of (a) neat epoxy and (b) 0.25 wt% CNT/epoxy composite aged at 55°C and 95% RH for various time lengths. (From Prolongo, S.G. et al., *Compos. Part A Appl. Sci. Manuf.*, 43, 2169–2175, 2012.)

Starkova et al. [133] have also measured the thermomechanical response of an epoxy and a CNT-reinforced epoxy after moisture ingression. With an increase in the water content, the storage modulus of both of the materials showed a continuous decrement. However, at the same moisture content, the CNT-reinforced epoxy exhibited a more improved storage modulus than the neat epoxy. In addition, due to water ingression, the storage modulus degrades at a faster rate for a neat epoxy than a nanocomposite. This improved water resistance of a CNT-based composite was attributed toward a lower amount of water molecules of type-I and, hence, an improved resistance of the nanocomposite toward plasticization.

Due to the hydrophobic nature of CNTs, it is rational to expect a better water resistance of a CNT-reinforced composite. However, the CNT/polymer interface plays a dominating role in deciding this aforesaid behavior. The existing literature are insufficient to draw any precise conclusion on this particular subject. Gkikas et al. [123] concluded that CNTs are ineffective in improving the water resistance of an epoxy, both from the diffusion rate and saturation content point of view. Whereas, Starkova et al. [133] claimed that CNT reinforcement helps in decreasing the water uptake rate, but the saturation water content remains unchanged. As per Prolongo et al. [131], both the diffusion rate and saturation water content were decreased due to CNT reinforcement in an epoxy. However, there are several underlying parameters which may play dominating roles on deciding such behavior, such as fabrication technique, CNT geometry and content, interfacial adhesion, and so on.

8.6.3 Ultraviolet and Other High-Energy Irradiation

Sometime structural materials are subjected to high-energy radiation. Ultraviolet radiation is one of the commonly encountered forms of an electromagnetic (EM) wave that has a wavelength 10–400 nm. This UV radiation has further been categorized as per an ISO-21348 standard.

The ozone layer of Earth acts as a filtering membrane, which absorbs almost 94% of the solar radiation and the remaining 6% reaches the surface of Earth, which is in the UV range [134]. Polymers are usually sensitive to UV radiation having a wavelength of 290–400 nm. This is because the energy of this UV radiation is somehow similar to the bond energy of most of the polymers. Hence, interaction of these UV rays with the molecular bonds of the polymer may result in the dissociation of these bonds. This eventually results in alteration in the properties of the polymer [135]. Similarly, other forms of EM waves like neutron and γ rays may be deleterious toward the structural health of the polymer and its composite used in nuclear applications [136].

The effect of UV radiation and e-beams on the structural integrity of a CNT-reinforced poly(methyl methacrylate) nanocomposite has been evaluated by Nafaji et al. [137] by analyzing the sample thickness and surface morphology. CNT addition in the polymer has been found to be beneficial toward improving its resistance against these radiations. As per their investigation, a network structure of CNTs was obtained in the case of a nanocomposite with 0.5% CNT. The reason has been attributed to the dissipation of the high-energy radiation through this continuous CNT network in the nanocomposite, which enhances its damage tolerance.

A detailed study on the durability of a CNT (3.5%) -reinforced epoxy nanocomposite has been done by Petersen et al. [138] by the help of weight loss, surface composition, and morphology. This study has been performed at different UV doses with a maximum value of 1089 MJ/m². Initially, a small increment in the weight of the samples was obtained due to moisture ingress from the ageing chamber. With the maximum UV dose (i.e., 1089 MJ/m²), the weight loss for the neat epoxy and CNT/epoxy nanocomposite was found to be 2.3% and 1%, respectively. The presence of CNT on the sample surface reduces the affinity of the material toward photodegradation. To understand the chemical changes on the surface of the samples due to UV exposure, FTIR–ATR spectra have been obtained for both the neat epoxy and nanocomposite, as shown in Figure 8.30. As the intensity of UV radiation was increased, the height of various bands continued to decrease, which are related to the structure of an epoxy, for example C–O band ($\sim$1,245 cm^{-1}) and benzene ring (1,508 cm^{-1}). At the same time, new bands also appeared at 1,620–1,740 cm^{-1}, corresponding to C=C and an aldehyde or ketone type of carbonyl groups. Extensive scission of the polymeric main chain due to the photodegradation behavior brought these changes in the FTIR spectra. Further, both of the bands at 1,508 and 1,726 cm^{-1} reached a plateau in the absorption intensity profiles at an UV intensity of 166 and 276 MJ/m² for the nanocomposite and neat epoxy, respectively. The dissipation of the UV energy through the conducting CNT network, the infiltration of the UV radiation by the electron ring of the CNT network, and the interaction of CNT with the free radicals can explain this behavior.

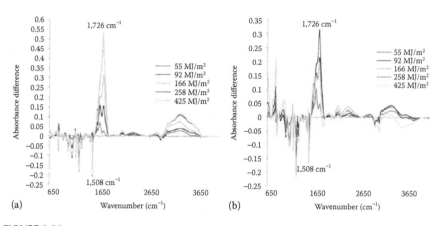

FIGURE 8.30
Difference in FTIR–ATR spectra obtained at different UV doses for (a) neat epoxy and (b) MWCNT/epoxy nanocomposite. (From Petersen, E.J. et al., *Carbon*, 69, 194–205, 2014.)

The morphological changes of the surface of the nanocomposite after the UV treatment of 1089 MJ/m² have been presented in Figure 8.31. A fairly smooth and flat surface of the nanocomposite can be noticed before the UV treatment. However, the surface morphology has been substantially changed after the UV treatment, making it relatively rough. This is due to the degradation of the top layer polymer leaving behind bare nanotubes.

Wohlleben et al. [139] have made a similar observation. They have investigated the UV durability of CNT-reinforced polyurethane (PU). The surface morphology of both a neat PU and CNT/PU nanocomposites before and after a UV treatment can be seen from Figure 8.32. In the case of a nanocomposite, the top surface layer polymer has been dissolved and the bare CNTs, which are resistant to such UV radiation, remained there unaffected. They have further reported the combined accelerated deleterious effect of UV treatment in association with environmental humidity. A UV conditioning at 50 ± 10% RH, results in 3 times faster removal of the polymer. Due to such a high rate of polymer dissolution, the CNT content on the top (10 nm) layer of the nanocomposite was found to be 72 ± 3%.

8.6.4 Effect of Atomic Oxygen

When atomic oxygen travelling with a high velocity collides with polymers, which basically contain hydrocarbon groups, various bonds are broken, resulting in loss of strength and stiffness of the polymer [140]. This case is very prominent in the case of space vehicles, which mostly travel at a speed of about 8 km/s. Impingement of the atomic oxygen with this high-speed vehicle results in a very high-energy flux (5 eV) at the surface. Thus, various bonds of the polymer are damaged. Erosion caused due to this atomic oxygen is far faster than that caused due to thermal atomic oxygen [141]. Initially, when a

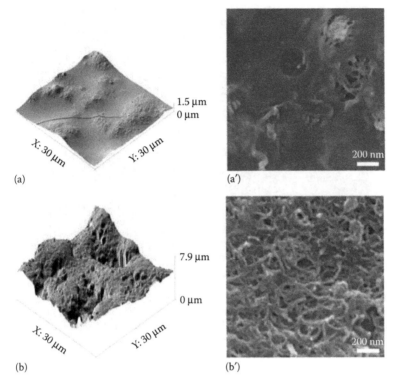

FIGURE 8.31
High-resolution atomic force microscopy and SEM images of CNT/epoxy nanocomposite (a, a') as fabricated and (b, b') after 1089 MJ/m² UV dose. (From Petersen, E.J. et al., *Carbon*, 69, 194–205, 2014.)

FRP composite is exposed to atomic oxygen, the rate of material removal is quite drastic. However, when the fibers are exposed to atomic oxygen, the rate is significantly reduced. In a similar way, if the polymer is modified with a CNT kind of nanofillers, which are resistant to atomic oxygen, the rate of erosion is expected to be reduced [142]. Figure 8.33 indicates the results obtained by Jiao et al. [143]. Both mass loss and erosion yield have been plotted against atomic oxygen (AO) fluence. As can be seen, the mass loss in the case of neat epoxy is quite fast. The addition of a CNT to an epoxy seems to improve this condition very efficiently. Both the CNT film and CNT/epoxy composite exhibit a similar extent of mass loss and erosion yield. Inertness of the graphitic CNT toward AO is responsible for such improved resistance against AO.

8.6.5 Exposure under Low Earth Orbit Space Environment

Strong, stiff, and yet light FRP composites are the present day material for space vehicular applications. Space vehicles are usually operated in a very

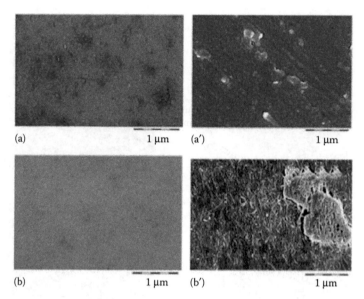

FIGURE 8.32
SEM images of (a) PU and (b) CNT/PU before UV exposure, (a') PU and (b') CNT/PU after
UV exposure (111 W m⁻² (300–400 nm)) of 8 weeks. (From Wohlleben, W. et al., *Nanoscale*, 5,
369–380, 2013.)

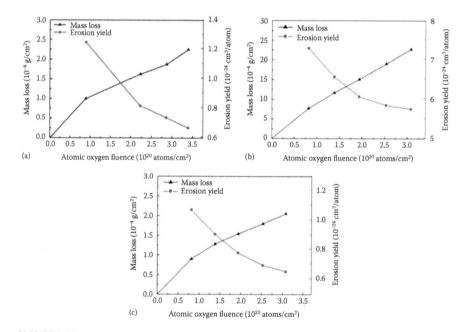

FIGURE 8.33
Mass loss and erosion yield of (a) CNT film, (b) neat epoxy, and (c) CNT/epoxy composite.
(From Jiao, L. et al., *Compos. Part A Appl. Sci. Manuf.*, 71, 116–125, 2015.)

complex environment which is composed of the atomic oxygen of high kinetic energy (~5 eV) with a flux of ~10^{14} atoms/cm^{2}s to 10^{15} atoms/cm^{2}s, high vacuum ~10^{-6} to 10^{-10} torr, highly fluctuating thermal cycles (–150°C to 150°C), high-energy UV radiation, electromagnetic waves, micrometeoroids, charged particles, and so on [144]. When a FRP composite is exposed to such an environment, its long-term durability must be well ensured under it. Several high-energy radiation and high-energy particles have been found to promote interfacial debonding. As explained earlier, AO causes surface erosion of the material. Repeated thermal cycling may result in fatigue crack generation and its subsequent growth. UV radiations also pose a threat on a FRP composite due to the photodegradation behavior of polymers [145,146]. In addition, some of the absorbed moisture and/or volatile gases, which may be entrapped in the composite during handling or fabrication, escape, leading to dimension change of the component, as well as contamination of nearby components [147,148]. Due to low Earth orbit (LEO) environmental exposure, the polymer gets preferentially affected. Hence, the flexural properties of a graphite/epoxy composite were degraded at a faster rate than that of other in-plane mechanical properties due to thermal cycling in a LEO environment [145]. Hence, to improve this condition, the matrix may be suitably modified so as to obtain better resistance against a LEO space environment. Hence, CNT-embedded polymers are the emerging materials for such applications.

Moon et al. [144] assessed the effect of CNT reinforcement on the LEO space environment durability. The surface morphology of a CFRP with and without a CNT after 20 h of exposure in a simulated LEO environment has been shown in Figure 8.34. Polymer from the control CFRP composite gets uniformly depleted due to LEO environmental exposure. However, in the case of a 0.7% CNT-embedded CFRP composite, erosion could take place until the subsurface CNTs get exposed to the environment. A relatively reduced rate of erosion could be noticed in the case of a CNT-embedded CFRP composite, due to inertness of a CNT toward this environment [149].

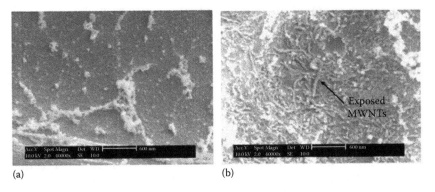

(a)   (b)

FIGURE 8.34
SEM image of the surface of (a) CFRP and (b) CFRP with 0.7% MWCNT after 20 h exposure to simulated LEO environment. (From Moon, J.-B. et al., *Compos. Part A Appl. Sci. Manuf.*, 42, 694–701, 2011.)

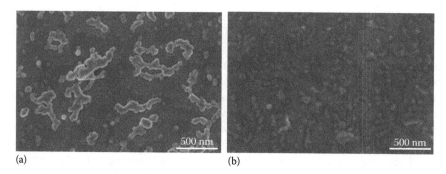

(a)   (b)

FIGURE 8.35

SEM image of the surface of nanocomposite with (a) pristine- and (b) silane-functionalized CNT after LEO environmental ageing. (From Jin, S.B. et al., *Compos. Sci. Technol.*, 91, 105, 2014.)

The strength of the control CFRP and CNT-embedded CFRP was reduced by 15.71% and 5.94%, respectively, due to this LEO environmental exposure.

Jin et al. [150] have studied the impact of silane functionalization of a CNT on the LEO environment durability of a CNT/epoxy composite. Thermal cycling of a neat epoxy, a pristine CNT/epoxy, and silane functionalized CNT/epoxy composites between −70°C and 100°C has been done in a LEO space environment. After 14 such repeated thermal cycles, the strength of a neat epoxy, a pristine CNT/epoxy, and silane functionalized CNT/epoxy composites were dropped by 10%, 8%, and 6%, respectively. A similar observation was also made when the mass loss was monitored. A pristine CNT/epoxy and silane functionalized CNT/epoxy composites resulted in a 5% and 9% lower mass loss, respectively, compared to a neat epoxy. As can be seen from the SEM images of both composites (Figure 8.35) after this conditioning, the extent of erosion seems lower for the silane functionalized CNT/epoxy composite.

8.6.6 Exposure to Electromagnetic and Microwave Radiation

Radiation of EM radiation from various electronic goods used for both civilian and military applications has an environmental threatening risk. Thus, from time-to-time environmental rules are getting stringent. However, serious efforts are being continuously made to develop new techniques and materials to restrict the EM waves going out of electronic devices and into the environment. Lightweight polymeric materials are the choice of people today for EM radiation absorption instead of old-fashioned heavy and rust prone materials based on ferrite. An improved impedance matching and good EM absorption potential are the required criteria for selecting a material for EM shielding [151]. In addition to extraordinary strength and stiffness, a CNT also exhibits good electrical as well as EM properties. This is due to the uninterrupted flow of electrons through the axis of a CNT, which is facilitated by the structure of a CNT (i.e., hollow cylindrical) [152].

Hence, rigorous research is currently done for assessing and improving the EM shielding behavior of CNT-based polymer materials.

Fan et al. [153] have investigated the effect of CNT content on the EM absorption (2–18 GHz) behavior of various polymers. In this frequency range, the loss tangent was determined and compared for nanocomposites with various CNT content. As per their investigation, the loss tangent of magnetic loss was quite negligible when compared with that of dielectric loss. Hence, the reduced microwave absorption behavior was thought to be evolved from the dielectric loss. With increase in CNT content, the reflectivity peak shifts toward a lower frequency as can be noticed from Figure 8.36a. At the same time, at a given frequency, the loss factor is higher for the composite with a higher CNT concentration, as evident from Figure 8.36b. Beyond a critical CNT content (4 wt%–8 wt%) in CNT/PET composite, there is a phase transformation converting the material from an insulator to a conductor (Figure 8.36b).

Rohini et al. [154] analyzed the effect of both pristine and functionalized CNT reinforcement on the EM shielding efficiency of polystyrene/PMMA blend. An improved EM shielding efficiency of more than 24 dB has been reported for both pristine and amino-functionalized CNT reinforcement. CNT gelation in the PMMA has been shown to be the contributory factor toward improved EM shielding efficiency of the composite.

Wang et al. [152] have reported a very well-organized, detailed study on the effect of CNT content on the overall permittivity, EM-shielding behavior, and conductivity of CNT/epoxy composites in the R band (26.5–40 GHz). All of these aforesaid properties were improved continuously due to an increase in CNT content up to 4%, as can be seen from Figure 8.37a. The superior intrinsic conductivity of nanotubes was accredited the cause of this improvement. However, at a CNT concentration of 5%, a sudden increase in these properties was noticed, due to the formation of CNT

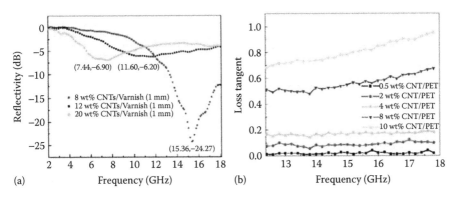

(a)   (b)

FIGURE 8.36

EM absorption properties of CNT-reinforced polymers, (a) reflectivity versus frequency of CNT/varnish composites and (b) Loss tangent versus frequency of CNT/PET composites at different CNT concentrations. (From Fan, Z. et al., *Mater. Sci. Eng. B*, 132, 85–89, 2006.)

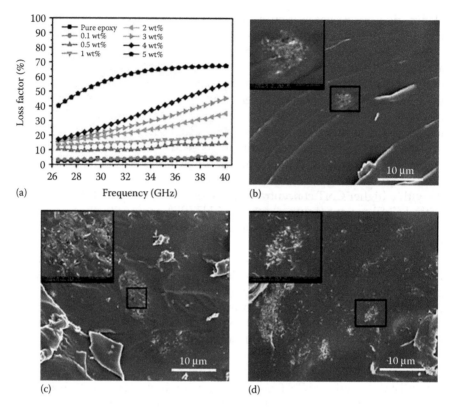

FIGURE 8.37
(a) Loss factor of neat and MWCNT/epoxy composites at various MWCNT loading, SEM images of MWCNT/epoxy composites at (b) 0.1 wt%, (c) 3 wt%, and (d) 5 wt% MWCNT (The insets on the upper-left of (b)–(d) show the regions in the black boxes with higher magnification). (From Wang, Z. and Zhao, G.-L., *J. Mater. Chem. C*, 2, 9406–9411, 2014.)

agglomerate and/or continuous network of nanotubes, as can be observed in Figure 8.37b and d. This essentially facilitates easy transportation of electrons and heat. Frequency has a vital role at a higher CNT content for deciding the absorbance (loss factor in Figure 8.37a) of the material. The interaction of the applied EM waves with the electrons of the nanotube makes dissipation of the radiation possible through it (in discrete or continuous network form). Therefore, the material exhibits a relatively higher EM absorption capacity.

Kong et al. [155] have investigated the EM absorption capacity of a hybrid polymer nanocomposite, where poly(dimethyl siloxane) was reinforced with both graphene oxide (GO) and CNTs. The CNTs were first grown on the surface of reduced graphene oxide (RGO) at 600°C. This CNT-deposited RGO was then dispersed in poly(dimethyl siloxane) for preparing the hybrid composite. The EM shielding efficiency of this nanocomposite was then evaluated in the X band. EM shielding efficiency (SE_T) can be defined from incident (P_I) and transmitted (P_T) power as expressed below:

$$SE_T = 10\log_{10}\left(\frac{P_I}{P_T}\right)$$

It can be seen from Figure 8.38 that the EM shielding efficiency increases with an increase in the CNT/graphene (CNT/G) concentration in the nanocomposite. With an increase in the CNT/G concentration from 2.5 wt% to 10 wt%, the EM shielding efficiency was increased by about 200%.

Jia et al. [156] have reported the EM absorption capacity of another hybrid composite based on a matrix of polymer blend (ethylene vinyl acetate and ultrahigh molecular weight polyethylene) and CNT. The composite with 20 wt% ethylene vinyl acetate and 7 wt% CNT yielded a 57.4 dB EM shielding capacity, which is substantially higher than many conducting polymers.

Several research groups are still working on polymer foam reinforced with CNT for developing lightweight EM shielding materials [157–160].

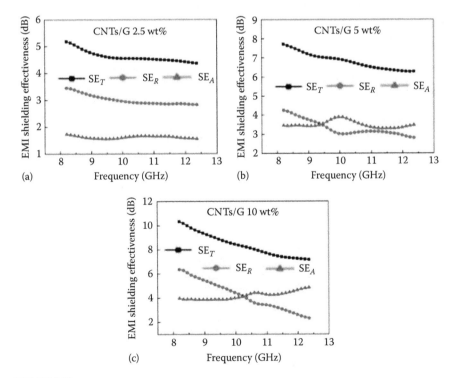

FIGURE 8.38
EM shielding efficiency of CNT/graphene (CNT/G)-reinforced poly(dimethyl siloxane) composite with (a) 2.5 wt%, (b) 5 wt%, and (c) 10 wt% CNT/G content. (From Kong, L. et al., *Carbon*, 73, 185–193, 2014.)

8.7 Summary

Even after two decades of devoted research in the field of CNT-embedded polymeric composites, the state-of-the-art potential of the CNTs in the final composite is not yet achieved. Hence, there is a stringent requirement for a better understanding of the polymer/CNT interaction at the interface, which holds the key for decisively achieving the goal. A stronger and stiffer CNT with a higher aspect ratio is desirable. The addition of a very minute quantity of CNT is capable of making the soft polymer reasonably stronger and stiffer. Judicious control over the CNT/polymer interface can even make the material significantly tougher. The processing technology has a direct impact on the dispersion of CNTs, which ultimately decides the interfacial area available for stress transfer. An urgent need persists to understand the chemistry at the composite interfaces and its alteration with time at various environments in unlocking the full potential of these materials. Unfortunately, these interfaces are difficult to assess as they have a dynamic character with time and environment. For a long-term reliable and durable structural application, the character and chemistry of the CNT/polymer interface at that particular application environment must be well ensured.

References

1. P. Compston, P.-Y. B. Jar, and P. Davies, Matrix effect on the static and dynamic interlaminar fracture toughness of glass-fibre marine composites, *Compos. Part B Eng.*, 29: 505–516, 1998. doi:10.1016/S1359-8368(98)00004-3.
2. H. Masaki, O. Shojiro, C.-G. Gustafson, and T. Keisuke, Effect of matrix resin on delamination fatigue crack growth in CFRP laminates, *Eng. Fract. Mech.*, 49: 35–47, 1994. doi:10.1016/0013-7944(94)90109-0.
3. R. P. Singh, M. Zhang, and D. Chan, Toughening of a brittle thermosetting polymer: Effects of reinforcement particle size and volume fraction, *J. Mater. Sci.*, 37: 781–788, 2002. doi:10.1023/A:1013844015493.
4. T. Kawaguchi and R. A. Pearson, The effect of particle-matrix adhesion on the mechanical behavior of glass filled epoxies. Part 2. A study on fracture toughness, *Polymer*, 44: 4239–4247, 2003. doi:10.1016/S0032-3861(03)00372-0.
5. A. C. Moloney, H. H. Kausch, T. Kaiser, and H. R. Beer, Parameters determining the strength and toughness of particulate filled epoxide resins, *J. Mater. Sci.*, 22: 381–393, 1987. doi:10.1007/BF01160743.
6. A. C. Garg and Y.-W. Mai, Failure mechanisms in toughened epoxy resins—A review, *Compos. Sci. Technol.*, 31: 179–223, 1988. doi:10.1016/0266-3538(88)90009-7.
7. R. A. Pearson, Toughening epoxies using rigid thermoplastic particles, in C. Keith Riew, and Anthony J. Kinloch (Eds.), *Toughened Plastics I*, vol. 233. Washington, DC: American Chemical Society, pp. 405–425, 1993.

8. D. A. Norman and R. E. Robertson, Rigid-particle toughening of glassy polymers, *Polymer*, 44: 2351–2362, 2003. doi:10.1016/S0032-3861(03)00084-3.
9. G. Keledi, J. Hári, and B. Pukánszky, Polymer nanocomposites: Structure, interaction, and functionality, *Nanoscale*, 4: 1919–1938, 2012. doi:10.1039/C2NR11442A.
10. S.-Y. Fu, X.-Q. Feng, B. Lauke, and Y.-W. Mai. Effects of particle size, particle/matrix interface adhesion and particle loading on mechanical properties of particulate-polymer composites, *Compos. Part B Eng.*, 39: 933–961, 2008. doi:10.1016/j.compositesb.2008.01.002.
11. S. T. Knauert, J. F. Douglas, and F. W. Starr, The effect of nanoparticle shape on polymer-nanocomposite rheology and tensile strength, *J. Polym. Sci. Part B Polym. Phys.*, 45: 1882–1897, 2007. doi:10.1002/polb.21176.
12. C. L. Wu, M. Q. Zhang, M. Z. Rong, and K. Friedrich, Tensile performance improvement of low nanoparticles filled-polypropylene composites, *Compos. Sci. Technol.*, 62: 1327–1340, 2002. doi:10.1016/S0266-3538(02)00079-9.
13. B. Fiedler, F. H. Gojny, M. H. G. Wichmann, M. C. M. Nolte, and K. Schulte, Fundamental aspects of nano-reinforced composites, *Compos. Sci. Technol.*, 66: 3115–3125, 2006. doi:10.1016/j.compscitech.2005.01.014.
14. P. K. Mallick and L. J. Broutman, Mechanical and fracture behaviour of glass bead filled epoxy composites, *Mater. Sci. Eng.*, 18: 63–73, 1975. doi:10.1016/0025-5416(75)90073-7.
15. D. Hull and T. W. Clyne, *An Introduction to Composite Materials*. New York: Cambridge University Press, 1996.
16. A. Kelly and W. R. Tyson, Tensile properties of fibre-reinforced metals: Copper/tungsten and copper/molybdenum, *J. Mech. Phys. Solids*, 13: 329–350, 1965. doi:10.1016/0022-5096(65)90035-9.
17. R. Balasubramaniam, *Callister's Materials Science and Engineering: Indian Adaptation (W/Cd)*. New Delhi: John Wiley & Sons, 2007.
18. J. N. Coleman, M. Cadek, R. Blake, V. Nicolosi, K. P. Ryan, C. Belton et al., High Performance nanotube-reinforced plastics: Understanding the mechanism of strength increase, *Adv. Funct. Mater.*, 14: 791–798, 2004. doi:10.1002/adfm.200305200.
19. H. L. Cox, The elasticity and strength of paper and other fibrous materials, *Br. J. Appl. Phys.*, 3: 72, 1952. doi:10.1088/0508-3443/3/3/302.
20. J. N. Coleman, U. Khan, W. J. Blau, and Y. K. Gun'ko, Small but strong: A review of the mechanical properties of carbon nanotube-polymer composites, *Carbon*, 44: 1624–1652, 2006. doi:10.1016/j.carbon.2006.02.038.
21. H. Krenchel, *Fibre Reinforcement: Theoretical and Practical Investigations of the Elasticity and Strength of Fibre-Reinforced Materials*. Copenhagen, Denmark: Akademisk Forlag, 1964.
22. J. C. H. Affdl and J. L. Kardos, The Halpin-Tsai equations: A review, *Polym. Eng. Sci.*, 16: 344–352, 1976. doi:10.1002/pen.760160512.
23. M. Mu and K. I. Winey, Improved load transfer in nanotube/polymer composites with increased polymer molecular weight, *J. Phys. Chem. C*, 111: 17923–17927, 2007. doi:10.1021/jp0715530.
24. J. Gao, M. A. Loi, E. J. F. de Carvalho, and M. C. dos Santos, Selective wrapping and supramolecular structures of polyfluorene-carbon nanotube hybrids, *ACS Nano.*, 5: 3993–3999, 2011. doi:10.1021/nn200564n.
25. W. Yi, A. Malkovskiy, Y. Xu, X.-Q. Wang, A. P. Sokolov, M. Lebron-Colon et al., Polymer conformation-assisted wrapping of single-walled carbon nanotube:

The impact of cis-vinylene linkage, *Polymer*, 51: 475–481, 2010. doi:10.1016/j.polymer.2009.11.052.

26. I. Kusner and S. Srebnik, Conformational behavior of semi-flexible polymers confined to a cylindrical surface, *Chem. Phys. Lett.*, 430: 84–88, 2006. doi:10.1016/j.cplett.2006.08.085.

27. S. S. Tallury and M. A. Pasquinelli, Molecular dynamics simulations of polymers with stiff backbones interacting with single-walled carbon nanotubes, *J. Phys. Chem. B*, 114: 9349–9355, 2010. doi:10.1021/jp101191j.

28. S. S. Tallury and M. A. Pasquinelli, Molecular dynamics simulations of flexible polymer chains wrapping single-walled carbon nanotubes, *J. Phys. Chem. B*, 114: 4122–4129, 2010. doi:10.1021/jp908001d.

29. M. Yang, V. Koutsos, and M. Zaiser, Interactions between polymers and carbon nanotubes: A molecular dynamics study, *J. Phys. Chem. B*, 109: 10009–10014, 2005. doi:10.1021/jp0442403.

30. B. McCarthy, J. N. Coleman, R. Czerw, A. B. Dalton, M. in het Panhuis, A. Maiti et al., A microscopic and spectroscopic study of interactions between carbon nanotubes and a conjugated polymer, *J. Phys. Chem. B*, 106: 2210–2216, 2002. doi:10.1021/jp013745f.

31. B. McCarthy, J. N. Coleman, S. A. Curran, A. B. Dalton, A. P. Davey, Z. Konya et al., Observation of site selective binding in a polymer nanotube composite, *J. Mater. Sci. Lett.*, 19: 2239–2241, 2000. doi:10.1023/A:1006776908183.

32. W. Ding, A. Eitan, F. T. Fisher, X. Chen, D. A. Dikin, R. Andrews et al., Direct observation of polymer sheathing in carbon nanotube-polycarbonate composites, *Nano. Lett.*, 3: 1593–1597, 2003. doi:10.1021/nl0345973.

33. A. H. Barber, S. R. Cohen, and H. D. Wagner, Measurement of carbon nanotube-polymer interfacial strength, *Appl. Phys. Lett.*, 82: 4140–4142, 2003. doi:10.1063/1.1579568.

34. S. W. Kim, T. Kim, Y. S. Kim, H. S. Choi, H. J. Lim, S. J. Yang et al., Surface modifications for the effective dispersion of carbon nanotubes in solvents and polymers, *Carbon*, 50: 3–33, 2012. doi:10.1016/j.carbon.2011.08.011.

35. T. Ramanathan, A. A. Abdala, S. Stankovich, D. A. Dikin, M. Herrera-Alonso, R. D. Piner et al., Functionalized graphene sheets for polymer nanocomposites, *Nat. Nanotechnol.*, 3: 327–331, 2008. doi:10.1038/nnano.2008.96.

36. J.-M. Yuan, Z.-F. Fan, X.-H. Chen, X.-H. Chen, Z.-J. Wu, and L.-P. He, Preparation of polystyrene-multiwalled carbon nanotube composites with individual-dispersed nanotubes and strong interfacial adhesion, *Polymer*, 50: 3285–3291, 2009. doi:10.1016/j.polymer.2009.04.065.

37. A. Eitan, K. Jiang, D. Dukes, R. Andrews, and L. S. Schadler, Surface modification of multiwalled carbon nanotubes: Toward the tailoring of the interface in polymer composites, *Chem. Mater.*, 15: 3198–3201, 2003. doi:10.1021/cm020975d.

38. J. Zhu, H. Peng, F. Rodriguez-Macias, J. L. Margrave, V. N. Khabashesku, A. M. Imam et al., Reinforcing epoxy polymer composites through covalent integration of functionalized nanotubes, *Adv. Funct. Mater.*, 14: 643–648, 2004. doi:10.1002/adfm.200305162.

39. Q. Zheng, Q. Xue, K. Yan, X. Gao, Q. Li, and L. Hao, Effect of chemisorption on the interfacial bonding characteristics of carbon nanotube-polymer composites, *Polymer*, 49: 800–808, 2008. doi:10.1016/j.polymer.2007.12.018.

40. F. Buffa, G. A. Abraham, B. P. Grady, and D. Resasco, Effect of nanotube function-alization on the properties of single-walled carbon nanotube/polyurethane com-posites, *J. Polym. Sci. Part B Polym. Phys.*, 45: 490–501, 2007. doi:10.1002/polb.21069.

41. J. Zhu, J. Kim, H. Peng, J. L. Margrave, V. N. Khabashesku, and E. V. Barrera, Improving the dispersion and integration of single-walled carbon nanotubes in epoxy composites through functionalization, *Nano Lett.*, 3: 1107–1113, 2003. doi:10.1021/nl0342489.

42. S. J. V. Frankland, A. Caglar, D. W. Brenner, and M. Griebel, Molecular simu-lation of the influence of chemical cross-links on the shear strength of carbon nanotube-polymer interfaces, *J. Phys. Chem. B*, 106: 3046–3048, 2002. doi:10.1021/jp015591+.

43. C. Velasco-Santos, A. L. Martínez-Hernández, F. T. Fisher, R. Ruoff, and V. M. Castaño, Improvement of thermal and mechanical properties of carbon nanotube composites through chemical functionalization, *Chem. Mater.*, 15: 4470–4475, 2003. doi:10.1021/cm034243c.

44. A. A. Koval'chuk, V. G. Shevchenko, A. N. Shchegolikhin, P. M. Nedorezova, A. N. Klyamkina, and A. M. Aladyshev, Effect of carbon nanotube functional-ization on the structural and mechanical properties of polypropylene/MWCNT composites, *Macromolecules*, 41: 7536–7542, 2008. doi:10.1021/ma801599q.

45. C.-M. Chang and Y.-L. Liu, Functionalization of multi-walled carbon nano-tubes with non-reactive polymers through an ozone-mediated process for the preparation of a wide range of high performance polymer/carbon nanotube composites, *Carbon*, 48: 1289–1297, 2010. doi:10.1016/j.carbon.2009.12.002.

46. H. C. Shim, Y. K. Kwak, C.-S. Han, and S. Kim, Enhancement of adhesion between carbon nanotubes and polymer substrates using microwave irradia-tion, *Scr. Mater.*, 61: 32–35, 2009. doi:10.1016/j.scriptamat.2009.02.060.

47. E. Moaseri, S. Hasanabadi, M. Maghrebi, and M. Baniadam, Improvements in fatigue life of amine-functionalized multi-walled carbon nanotube-reinforced epoxy composites: Effect of functionalization degree and microwave-assisted procuring, *J. Compos. Mater.*, 0021998314541306, 2014. doi:10.1177/0021998314541306.

48. T. Ramanathan, F. T. Fisher, R. S. Ruoff, and L. C. Brinson, Amino-functionalized carbon nanotubes for binding to polymers and biological systems, *Chem. Mater.*, 17: 1290–1295, 2005. doi:10.1021/cm048357f.

49. L. Sun, G. L. Warren, J. Y. O'Reilly, W. N. Everett, S. M. Lee, D. Davis et al., Mechanical properties of surface-functionalized SWCNT/epoxy composites, *Carbon*, 46: 320–328, 2008. doi:10.1016/j.carbon.2007.11.051.

50. H. Yu, Y. Jin, F. Peng, H. Wang, and J. Yang, Kinetically controlled side-wall functionalization of carbon nanotubes by nitric acid oxidation, *J. Phys. Chem. C*, 112: 6758–6763, 2008. doi:10.1021/jp711975a.

51. H. Florian and K. S. Gojny, Functionalisation effect on the thermo-mechanical behaviour of multi-wall carbon nanotube/epoxy-composites, *Compos. Sci. Technol.*, 2303–2308, 2004. doi:10.1016/j.compscitech.2004.01.024.

52. J. Shen, W. Huang, L. Wu, Y. Hu, and M. Ye, The reinforcement role of dif-ferent amino-functionalized multi-walled carbon nanotubes in epoxy nanocomposites, *Compos. Sci. Technol.*, 67: 3041–3050, 2007. doi:10.1016/j.compscitech.2007.04.025.

53. J. Li, Z. Fang, L. Tong, A. Gu, and F. Liu, Improving dispersion of multiwalled carbon nanotubes in polyamide 6 composites through amino-functionalization, *J. Appl. Polym. Sci.*, 106: 2898–2906, 2007. doi:10.1002/app.24599.

54. F. H. Gojny, M. H. G. Wichmann, U. Köpke, B. Fiedler, and K. Schulte, Carbon nanotube-reinforced epoxy-composites: Enhanced stiffness and fracture toughness at low nanotube content, *Compos. Sci. Technol.*, 64: 2363–2371, 2004. doi:10.1016/j.compscitech.2004.04.002.

55. F.-L. Jin, C.-J. Ma, and S.-J. Park, Thermal and mechanical interfacial properties of epoxy composites based on functionalized carbon nanotubes, *Mater. Sci. Eng. A*, 528: 8517–8522, 2011. doi:10.1016/j.msea.2011.08.054.

56. S. A. Sydlik, J.-H. Lee, J. J. Walish, E. L. Thomas, and T. M. Swager, Epoxy functionalized multi-walled carbon nanotubes for improved adhesives, *Carbon*, 59: 109–120, 2013. doi:10.1016/j.carbon.2013.02.061.

57. M. K. Bayazit and K. S. Coleman, Ester-functionalized single-walled carbon nanotubes via addition of haloformates, *J. Mater. Sci.*, 49: 5190–5198, 2014. doi:10.1007/s10853-014-8227-y.

58. P. C. Ma, J.-K. Kim, and B. Z. Tang, Functionalization of carbon nanotubes using a silane coupling agent, *Carbon*, 44: 3232–3238, 2006. doi:10.1016/j.carbon.2006.06.032.

59. D. Vennerberg, Z. Rueger, and M. R. Kessler, Effect of silane structure on the properties of silanized multiwalled carbon nanotube-epoxy nanocomposites, *Polymer*, 55: 1854–1865, 2014. doi:10.1016/j.polymer.2014.02.018.

60. J. Kathi, K.-Y. Rhee, and J. H. Lee, Effect of chemical functionalization of multi-walled carbon nanotubes with 3-aminopropyltriethoxysilane on mechanical and morphological properties of epoxy nanocomposites, *Compos. Part A Appl. Sci. Manuf.*, 40: 800–809, 2009. doi:10.1016/j.compositesa.2009.04.001.

61. M. T. Kim, K. Y. Rhee, J. H. Lee, D. Hui, and A. K. T. Lau, Property enhancement of a carbon fiber/epoxy composite by using carbon nanotubes, *Compos. Part B Eng.*, 42: 1257–1261, 2011. doi:10.1016/j.compositesb.2011.02.005.

62. X. Li, Y. Qin, S. T. Picraux, and Z.-X. Guo, Noncovalent assembly of carbon nanotube-inorganic hybrids, *J. Mater. Chem.*, 21: 7527–7547, 2011. doi:10.1039/C1JM10516G.

63. X. Gong, J. Liu, S. Baskaran, R. D. Voise, and J. S. Young, Surfactant-assisted processing of carbon nanotube/polymer composites, *Chem. Mater.*, 12: 1049–1052, 2000. doi:10.1021/cm9906396.

64. A. Mandal and A. K. Nandi, Noncovalent functionalization of multiwalled carbon nanotube by a polythiophene-based compatibilizer: Reinforcement and conductivity improvement in poly(vinylidene fluoride) films, *J. Phys. Chem. C*, 116: 9360–9371, 2012. doi:10.1021/jp302027y.

65. A. Star, J. F. Stoddart, D. Steuerman, M. Diehl, A. Boukai, E. W. Wong et al., Preparation and properties of polymer-wrapped single-walled carbon nanotubes, *Angew. Chem. Int. Ed.*, 40: 1721–1725, 2001. doi:10.1002/1521-3773(20010504)40:9<1721::AID-ANIE17210>3.0.CO;2-F.

66. M. J. Green, Analysis and measurement of carbon nanotube dispersions: Nanodispersion versus macrodispersion, *Polym. Int.*, 59: 1319–1322, 2010. doi:10.1002/pi.2878.

67. Z. Jin, K. P. Pramoda, G. Xu, and S. H. Goh, Dynamic mechanical behavior of melt-processed multi-walled carbon nanotube/poly(methyl methacrylate) composites, *Chem. Phys. Lett.*, 337: 43–47, 2001. doi:10.1016/S0009-2614(01)00186-5.

68. K.-K. Koo, T. Inoue, and K. Miyasaka, Toughened plastics consisting of brittle particles and ductile matrix, *Polym. Eng. Sci.*, 25: 741–746, 1985. doi:10.1002/pen.760251203.

69. M. Marić and C. W. Macosko, Improving polymer blend dispersion in minimixers, *Polym. Eng. Sci.*, 41: 118–130, 2001. doi:10.1002/pen.10714.

70. B. Yang, M. Sato, T. Kuriyama, and T. Inoue, Improvement of a gram-scale mixer for polymer blending, *J. Appl. Polym. Sci.*, 99: 1–5, 2006. doi:10.1002/app.20662.

71. G. J. Listner and A. J. Sampson, Single screw extruder. US3496603 A, 1970.

72. K. Wakabayashi, C. Pierre, D. A. Dikin, R. S. Ruoff, T. Ramanathan, L. C. Brinson et al., Polymer-graphite nanocomposites: Effective dispersion and major property enhancement via solid-state shear pulverization, *Macromolecules*, 41: 1905–1908, 2008.

73. S. Baumgartner, Twin screw extruder. US5000900 A, 1991.

74. E. T. Thostenson and T.-W. Chou, Aligned multi-walled carbon nanotube-reinforced composites: Processing and mechanical characterization, *J. Phys. Appl. Phys.*, 35: L77, 2002. doi:10.1088/0022-3727/35/16/103.

75. W. Lertwimolnun and B. Vergnes, Effect of processing conditions on the formation of polypropylene/organoclay nanocomposites in a twin screw extruder, *Polym. Eng. Sci.*, 46: 314–323, 2006. doi:10.1002/pen.20458.

76. S. J. Kim and T. H. Kwon, Enhancement of mixing performance of single-screw extrusion processes via chaotic flows: Part I. Basic concepts and experimental study, *Adv. Polym. Technol.*, 15: 41–54, 1996. doi:10.1002/(SICI)1098-2329(199621)15:1<41::AID-ADV4>3.0.CO;2-K.

77. L. Jin, C. Bower, and O. Zhou, Alignment of carbon nanotubes in a polymer matrix by mechanical stretching, *Appl. Phys. Lett.*, 73: 1197–1199, 1998. doi:10.1063/1.122125.

78. M. S. P. Shaffer and A. H. Windle, Fabrication and characterization of carbon nanotube/poly(vinyl alcohol) composites, *Adv. Mater.*, 11: 937–941, 1999. doi:10.1002/(SICI)1521-4095(199908)11:11<937::AID-ADMA937>3.0.CO;2-9.

79. D. Qian, E. C. Dickey, R. Andrews, and T. Rantell, Load transfer and deformation mechanisms in carbon nanotube-polystyrene composites, *Appl. Phys. Lett.*, 76: 2868–2870, 2000. doi:10.1063/1.126500.

80. B. Safadi, R. Andrews, and E. A. Grulke, Multiwalled carbon nanotube polymer composites: Synthesis and characterization of thin films, *J. Appl. Polym. Sci.*, 84: 2660–2669, 2002. doi:10.1002/app.10436.

81. S. L. Ruan, P. Gao, X. G. Yang, and T. X. Yu, Toughening high performance ultrahigh molecular weight polyethylene using multiwalled carbon nanotubes, *Polymer*, 44: 5643–5654, 2003. doi:10.1016/S0032-3861(03)00628-1.

82. K. L. Lu, R. M. Lago, Y. K. Chen, M. L. H. Green, P. J. F. Harris, and S. C. Tsang, Mechanical damage of carbon nanotubes by ultrasound, *Carbon*, 34: 814–816, 1996. doi:10.1016/0008-6223(96)89470-X.

83. Y. S. Song and J. R. Youn, Influence of dispersion states of carbon nanotubes on physical properties of epoxy nanocomposites, *Carbon*, 43: 1378–1385, 2005. doi:10.1016/j.carbon.2005.01.007.

84. J. Sandler, M. S. P. Shaffer, T. Prasse, W. Bauhofer, K. Schulte, and A. H. Windle, Development of a dispersion process for carbon nanotubes in an epoxy matrix and the resulting electrical properties, *Polymer*, 40: 5967–5971, 1999. doi:10.1016/S0032-3861(99)00166-4.

85. C. F. Schmid and D. J. Klingenberg, Mechanical flocculation in flowing fiber suspensions, *Phys. Rev. Lett.*, 84: 290–293, 2000. doi:10.1103/PhysRevLett.84.290.

86. A. Allaoui, S. Bai, H. M. Cheng, and J. B. Bai, Mechanical and electrical properties of a MWNT/epoxy composite, *Compos. Sci. Technol.*, 62: 1993–1998, 2002. doi:10.1016/S0266-3538(02)00129-X.

87. X. L. Xie, K. Aloys, X. P. Zhou, and F. D. Zeng, Ultrahigh molecular mass polyethylene/carbon nanotube composites crystallization and melting properties, *J. Therm. Anal. Calorim.*, 74: 317–323, 2003. doi:10.1023/A:1026362727368.

88. Z. Jin, K. P. Pramoda, S. H. Goh, and G. Xu, Poly(vinylidene fluoride)-assisted melt-blending of multi-walled carbon nanotube/poly(methyl methacrylate) composites, *Mater. Res. Bull.*, 37: 271–278, 2002. doi:10.1016/S0025-5408(01)00775-9.

89. P.-C. Ma, N. A. Siddiqui, G. Marom, and J.-K. Kim, Dispersion and functionalization of carbon nanotubes for polymer-based nanocomposites: A review, *Compos. Part A Appl. Sci. Manuf.*, 41: 1345–1367, 2010. doi:10.1016/j.compositesa.2010.07.003.

90. Z. Shen, S. Bateman, D. Y. Wu, P. McMahon, M. Dell'Olio, and J. Gotama, The effects of carbon nanotubes on mechanical and thermal properties of woven glass fibre reinforced polyamide-6 nanocomposites, *Compos. Sci. Technol.*, 69: 239–244, 2009. doi:10.1016/j.compscitech.2008.10.017.

91. A. M. Díez-Pascual, B. Ashrafi, M. Naffakh, J. M. González-Domínguez, A. Johnston, B. Simard et al., Influence of carbon nanotubes on the thermal, electrical and mechanical properties of poly(ether ether ketone)/glass fiber laminates, *Carbon*, 49: 2817–2833, 2011. doi:10.1016/j.carbon.2011.03.011.

92. A. M. Díez-Pascual, J. M. González-Domínguez, M. Teresa Martínez, and M. A. Gómez-Fatou, Poly(ether ether ketone)-based hierarchical composites for tribological applications, *Chem. Eng. J.*, 218: 285–294, 2013. doi:10.1016/j.cej.2012.12.056.

93. A. M. Díez-Pascual, M. Naffakh, C. Marco, M. A. Gómez-Fatou, and G. J. Ellis, Multiscale fiber-reinforced thermoplastic composites incorporating carbon nanotubes: A review, *Curr. Opin. Solid State Mater. Sci.*, 18: 62–80, 2014. doi:10.1016/j.cossms.2013.06.003.

94. A. M. Díez-Pascual, M. Naffakh, J. M. González-Domínguez, A. Ansón, Y. Martínez-Rubi, M. T. Martínez et al., High performance PEEK/carbon nanotube composites compatibilized with polysulfones-I. Structure and thermal properties, *Carbon*, 48: 3485–3499, 2010. doi:10.1016/j.carbon.2010.05.046.

95. Z. Fan, K.-T. Hsiao, and S. G. Advani, Experimental investigation of dispersion during flow of multi-walled carbon nanotube/polymer suspension in fibrous porous media, *Carbon*, 42: 871–876, 2004. doi:10.1016/j.carbon.2004.01.067.

96. F. H. Gojny, M. H. G. Wichmann, B. Fiedler, W. Bauhofer, and K. Schulte, Influence of nano-modification on the mechanical and electrical properties of conventional fibre-reinforced composites, *Compos. Part A Appl. Sci. Manuf.*, 36: 1525–1535, 2005. doi:10.1016/j.compositesa.2005.02.007.

97. L. Böger, M. H. G. Wichmann, L. O. Meyer, and K. Schulte, Load and health monitoring in glass fibre reinforced composites with an electrically conductive nanocomposite epoxy matrix, *Compos. Sci. Technol.*, 68: 1886–1894, 2008. doi:10.1016/j.compscitech.2008.01.001.

98. E. T. Thostenson, J. J. Gangloff, C. Li, and J.-H. Byun, Electrical anisotropy in multiscale nanotube/fiber hybrid composites, *Appl. Phys. Lett.*, 95, 2009. doi:10.1063/1.3202788.

99. D. W. Y. Wong, L. Lin, P. T. McGrail, T. Peijs, and P. J. Hogg, Improved fracture toughness of carbon fibre/epoxy composite laminates using dissolvable thermoplastic fibres, *Compos. Part A Appl. Sci. Manuf.*, 41: 759–767, 2010. doi:10.1016/j.compositesa.2010.02.008.

100. E. F. Reia da Costa, A. A. Skordos, I. K. Partridge, and A. Rezai, RTM processing and electrical performance of carbon nanotube modified epoxy/fibre composites, *Compos. Part A: Appl. Sci. Manuf.*, 43: 593–602, 2012. doi:10.1016/j.compositesa.2011.12.019.

101. F. H. Gojny, J. Nastalczyk, Z. Roslaniec, and K. Schulte, Surface modified multiwalled carbon nanotubes in CNT/epoxy-composites, *Chem. Phys. Lett.*, 370: 820–824, 2003. doi:10.1016/S0009-2614(03)00187-8.

102. F. H. Gojny and K. Schulte, Functionalisation effect on the thermo-mechanical behaviour of multi-wall carbon nanotube/epoxy-composites, *Compos. Sci. Technol.*, 64: 2303–2308, 2004. doi:10.1016/j.compscitech.2004.01.024.

103. E. Bekyarova, E. T. Thostenson, A. Yu, H. Kim, J. Gao, J. Tang et al., Multiscale carbon nanotube-carbon fiber reinforcement for advanced epoxy composites, *Langmuir*, 23: 3970–3974, 2007. doi:10.1021/la062743p.

104. M. H. G. Wichmann, J. Sumfleth, F. H. Gojny, M. Quaresimin, B. Fiedler, and K. Schulte, Glass-fibre-reinforced composites with enhanced mechanical and electrical properties: Benefits and limitations of a nanoparticle modified matrix, *Eng. Fract. Mech.*, 73: 2346–2359, 2006. doi:10.1016/j.engfracmech.2006.05.015.

105. K.-T. Hsiao, J. Alms, and S. G. Advani, Use of epoxy/multiwalled carbon nanotubes as adhesives to join graphite fibre reinforced polymer composites, *Nanotechnology*, 14: 791, 2003. doi:10.1088/0957-4484/14/7/316.

106. S. A. Meguid and Y. Sun, On the tensile and shear strength of nano-reinforced composite interfaces, *Mater. Des.*, 25: 289–296, 2004. doi:10.1016/j.matdes.2003.10.018.

107. K. C. Jajam, M. M. Rahman, M. V. Hosur, and H. V. Tippur, Fracture behavior of epoxy nanocomposites modified with polyol diluent and amino-functionalized multi-walled carbon nanotubes: A loading rate study, *Compos. Part A Appl. Sci. Manuf.*, 59: 57–69, 2014. doi:10.1016/j.compositesa.2013.12.014.

108. F. H. Gojny, M. H. G. Wichmann, B. Fiedler, and K. Schulte, Influence of different carbon nanotubes on the mechanical properties of epoxy matrix composites: A comparative study, *Compos. Sci. Technol.*, 65: 2300–2313, 2005. doi:10.1016/j.compscitech.2005.04.021.

109. M. J. Shukla, D. S. Kumar, D. K. Rathore, R. K. Prusty, and B. C. Ray, An assessment of flexural performance of liquid nitrogen conditioned glass/epoxy composites with multiwalled carbon naotube, *J. Compos. Mater.*, 0021998315615648, 2015. doi:10.1177/0021998315615648.

110. M. Li, Y. Gu, Y. Liu, Y. Li, and Z. Zhang, Interfacial improvement of carbon fiber/epoxy composites using a simple process for depositing commercially functionalized carbon nanotubes on the fibers, *Carbon*, 52: 109–121, 2013. doi:10.1016/j.carbon.2012.09.011.

111. S. Sethi and B. C. Ray, An assessment of mechanical behavior and fractography study of glass/epoxy composites at different temperatures and loading speeds, *Mater. Des.*, 64: 160–165, 2014. doi:10.1016/j.matdes.2014.07.017.

112. S. Sethi, D. K. Rathore, and B. C. Ray, Effects of temperature and loading speed on interface-dominated strength in fibre/polymer composites: An evaluation for in-situ environment, *Mater. Des.*, 65: 617–626, 2015. doi:10.1016/j.matdes.2014.09.053.

113. M. J. Shukla, D. S. Kumar, K. K. Mahato, D. K. Rathore, R. K. Prusty, and B. C. Ray, A comparative study of the mechanical performance of glass and glass/carbon hybrid polymer composites at different temperature environments, *IOP Conf. Ser. Mater. Sci. Eng.*, 75: 012002, 2015. doi:10.1088/1757-899X/75/1/012002.

114. B. C. Ray, Thermal shock on interfacial adhesion of thermally conditioned glass fiber/epoxy composites, *Mater. Lett.*, 58: 2175–2177, 2004. doi:10.1016/j.matlet.2004.01.035.

115. B. C. Ray, Freeze-thaw response of glass-polyester composites at different loading rates, *J. Reinf. Plast. Compos.*, 24: 1771–1776, 2005.

116. B. C. Ray, Thermal shock and thermal fatigue on delamination of glass-fiber-reinforced polymeric composites, *J. Reinf. Plast. Compos.*, 24: 111–116, 2005. doi:10.1177/0731684405042953.

117. Z.-K. Chen, J.-P. Yang, Q.-Q. Ni, S.-Y. Fu, and Y.-G. Huang, Reinforcement of epoxy resins with multi-walled carbon nanotubes for enhancing cryogenic mechanical properties, *Polymer*, 50: 4753–4759, 2009. doi:10.1016/j.polymer.2009.08.001.

118. C. J. Huang, S. Y. Fu, Y. H. Zhang, B. Lauke, L. F. Li, and L. Ye, Cryogenic properties of SiO_2/epoxy nanocomposites, *Cryogenics*, 45: 450–454, 2005. doi:10.1016/j.cryogenics.2005.03.003.

119. F. Y. Wu and H. M. Cheng, Structure and thermal expansion of multi-walled carbon nanotubes before and after high temperature treatment, *J. Phys. Appl. Phys.*, 38: 4302, 2005. doi:10.1088/0022-3727/38/24/006.

120. K. K. Chawla, *Composite Materials: Science and Engineering*. New York: Springer, 2012.

121. R. K. Prusty, D. K. Rathore, M. J. Shukla, and B. C. Ray, Flexural behaviour of CNT-filled glass/epoxy composites in an in-situ environment emphasizing temperature variation. *Compos. Part B Eng.*, 83: 166–174, 2015. doi:10.1016/j.compositesb.2015.08.035.

122. D. K. Rathore, R. K. Prusty, D. S. Kumar, and B. C. Ray, Mechanical performance of CNT-filled glass fiber/epoxy composite in in-situ elevated temperature environments emphasizing the role of CNT content, *Compos. Part A Appl. Sci. Manuf.*, 84: 364–376, 2016. doi:10.1016/j.compositesa.2016.02.020.

123. G. Gkikas, D.-D. Douka, N.-M. Barkoula, and A. S. Paipetis, Nano-enhanced composite materials under thermal shock and environmental degradation: A durability study, *Compos. Part B Eng.*, 70: 206–214, 2015. doi:10.1016/j.compositesb.2014.11.008.

124. J. Zhou and J. P. Lucas, The effects of a water environment on anomalous absorption behavior in graphite/epoxy composites, *Compos. Sci. Technol.*, 53: 57–64, 1995. doi:10.1016/0266-3538(94)00078-6.

125. J. Zhou and J. P. Lucas, Hygrothermal effects of epoxy resin. Part I: The nature of water in epoxy, *Polymer*, 40: 5505–5512, 1999. doi:10.1016/S0032-3861(98)00790-3.

126. J. Zhou and J. P. Lucas, Hygrothermal effects of epoxy resin. Part II: Variations of glass transition temperature, *Polymer*, 40: 5513–5522, 1999. doi:10.1016/S0032-3861(98)00791-5.

127. Y. C. Lin and X. Chen, Investigation of moisture diffusion in epoxy system: Experiments and molecular dynamics simulations, *Chem. Phys. Lett.*, 412: 322–326, 2005. doi:10.1016/j.cplett.2005.07.022.

128. Y. C. Lin and X. Chen, Moisture sorption-desorption-resorption characteristics and its effect on the mechanical behavior of the epoxy system, *Polymer*, 46: 11994–12003, 2005. doi:10.1016/j.polymer.2005.10.002.

129. S. Popineau, C. Rondeau-Mouro, C. Sulpice-Gaillet, and M. E. R. Shanahan, Free/bound water absorption in an epoxy adhesive, *Polymer*, 46: 10733–10740, 2005. doi:10.1016/j.polymer.2005.09.008.

130. C. Wu and W. Xu, Atomistic simulation study of absorbed water influence on structure and properties of crosslinked epoxy resin, *Polymer*, 48: 5440–5448, 2007. doi:10.1016/j.polymer.2007.06.038.

131. S. G. Prolongo, M. R. Gude, and A. Ureña, Water uptake of epoxy composites reinforced with carbon nanofillers, *Compos. Part A Appl. Sci. Manuf.*, 43: 2169–2175, 2012. doi:10.1016/j.compositesa.2012.07.014.

132. S. G. Prolongo, M. Campo, M. R. Gude, R. Chaos-Morán, and A. Ureña, Thermophysical characterisation of epoxy resin reinforced by amino-functionalized carbon nanofibers, *Compos. Sci. Technol.*, 69: 349–357, 2009. doi:10.1016/j.compscitech.2008.10.018.

133. O. Starkova, S. T. Buschhorn, E. Mannov, K. Schulte, and A. Aniskevich, Water transport in epoxy/MWCNT composites, *Eur. Polym. J.*, 49: 2138–2148, 2013. doi:10.1016/j.eurpolymj.2013.05.010.

134. D. Feldman, Polymer weathering: Photo-oxidation, *J. Polym. Environ.*, 10: 163–173, 2002. doi:10.1023/A:1021148205366.

135. T.-C. Nguyen, Y. Bai, X.-L. Zhao, and R. Al-Mahaidi, Effects of ultraviolet radiation and associated elevated temperature on mechanical performance of steel/CFRP double strap joints, *Compos. Struct.*, 94: 3563–3573, 2012. doi:10.1016/j.compstruct.2012.05.036.

136. S. Blazewicz, J. Piekarczyk, J. Chlopek, J. Blocki, J. Michalowski, M. Stodulski et al., Effect of neutron irradiation on the mechanical properties of graphite fiber-based composites, *Carbon*, 40: 721–727, 2002. doi:10.1016/S0008-6223(01)00179-8.

137. E. Najafi and K. Shin, Radiation resistant polymer-carbon nanotube nanocomposite thin films, *Colloids Surf. Physicochem. Eng. Asp.*, 257–258: 333–337, 2005. doi:10.1016/j.colsurfa.2004.10.076.

138. E. J. Petersen, T. Lam, J. M. Gorham, K. C. Scott, C. J. Long, D. Stanley et al., Methods to assess the impact of UV irradiation on the surface chemistry and structure of multiwall carbon nanotube epoxy nanocomposites, *Carbon*, 69: 194–205, 2014. doi:10.1016/j.carbon.2013.12.016.

139. W. Wohlleben, M. W. Meier, S. Vogel, R. Landsiedel, G. Cox, S. Hirth et al., Elastic CNT-polyurethane nanocomposite: Synthesis, performance and assessment of fragments released during use, *Nanoscale*, 5: 369–380, 2013.

140. B. A. Banks, S. K. Rutledge, and J. A. Brady, The NASA atomic oxygen effects test program, 1988.

141. A. Paillous and C. Pailler, Degradation of multiply polymer-matrix composites induced by space environment, *Composites*, 25: 287–295, 1994. doi:10.1016/0010-4361(94)90221-6.

142. H. E. Misak, V. Sabelkin, S. Mall, and P. E. Kladitis, Thermal fatigue and hypothermal atomic oxygen exposure behavior of carbon nanotube wire, *Carbon*, 57: 42–49, 2013. doi:10.1016/j.carbon.2013.01.028.

143. L. Jiao, Y. Gu, S. Wang, Z. Yang, H. Wang, Q. Li et al., Atomic oxygen exposure behaviors of CVD-grown carbon nanotube film and its polymer composite film, *Compos. Part A Appl. Sci. Manuf.*, 71: 116–125, 2015. doi:10.1016/j.compositesa.2015.01.008.

144. J.-B. Moon, M.-G. Kim, C.-G. Kim, and S. Bhowmik, Improvement of tensile properties of CFRP composites under LEO space environment by applying MWNTs and thin-ply, *Compos. Part A Appl. Sci. Manuf.*, 42: 694–701, 2011. doi:10.1016/j.compositesa.2011.02.011.

145. K.-B. Shin, C.-G. Kim, C.-S. Hong, and H.-H. Lee, Prediction of failure thermal cycles in graphite/epoxy composite materials under simulated low Earth orbit environments, *Compos. Part B Eng.*, 31: 223–235, 2000. doi:10.1016/S1359-8368(99)00073-6.

146. F. Awaja, J. B. Moon, M. Gilbert, S. Zhang, C. G. Kim, and P. J. Pigram, Surface molecular degradation of selected high performance polymer composites under low Earth orbit environmental conditions, *Polym. Degrad. Stab.*, 96: 1301–1309, 2011. doi:10.1016/j.polymdegradstab.2011.04.001.

147. D. K. Felbeck, Toughened graphite-epoxy composites exposed in near-Earth orbit for 5.8 years, *J. Spacecr. Rockets*, 32: 317–323, 1995. doi:10.2514/3.26612.

148. J.-H. Han and C.-G. Kim, Low Earth orbit space environment simulation and its effects on graphite/epoxy composites, *Compos. Struct.*, 72: 218–226, 2006. doi:10.1016/j.compstruct.2004.11.007.

149. M. R. Reddy, Effect of low earth orbit atomic oxygen on spacecraft materials, *J. Mater. Sci.*, 30: 281–307, 1995. doi:10.1007/BF00354389.

150. S. B. Jin, G. S. Son, Y. H. Kim, and C. G. Kim, Enhanced durability of silanized multi-walled carbon nanotube/epoxy nanocomposites under simulated low Earth orbit space environment, *Compos. Sci. Technol.*, 91: 105, 2014. doi:10.1016/j.compscitech.2013.12.018.

151. R. Kumaran, S. D. Kumar, N. Balasubramanian, M. Alagar, V. Subramanian, and K. Dinakaran, Enhanced electromagnetic interference shielding in a Au-MWCNT composite nanostructure dispersed PVDF thin films, *J. Phys. Chem. C*, 120: 13771–13778, 2016. doi:10.1021/acs.jpcc.6b01333.

152. Z. Wang and G.-L. Zhao, Electromagnetic wave absorption of multi-walled carbon nanotube-epoxy composites in the R band, *J. Mater. Chem. C*, 2: 9406–9411, 2014.

153. Z. Fan, G. Luo, Z. Zhang, L. Zhou, and F. Wei, Electromagnetic and microwave absorbing properties of multi-walled carbon nanotubes/polymer composites, *Mater. Sci. Eng. B*, 132: 85–89, 2006. doi:10.1016/j.mseb.2006.02.045.

154. R. Rohini and S. Bose, Electromagnetic interference shielding materials derived from gelation of multiwall carbon nanotubes in polystyrene/poly(methyl methacrylate) blends, *ACS Appl. Mater. Interfaces*, 6: 11302–11310, 2014. doi:10.1021/am502641h.

155. L. Kong, X. Yin, X. Yuan, Y. Zhang, X. Liu, L. Cheng et al., Electromagnetic wave absorption properties of graphene modified with carbon nanotube/poly(dimethyl siloxane) composites, *Carbon*, 73: 185–193, 2014. doi:10.1016/j.carbon.2014.02.054.

156. L.-C. Jia, D.-X. Yan, C.-H. Cui, X. Ji, and Z.-M. Li, A unique double percolated polymer composite for highly efficient electromagnetic interference shielding, *Macromol. Mater. Eng.*, 301: 1232–1241, 2016. doi:10.1002/mame.201600145.

157. Y. Xu, Y. Li, W. Hua, A. Zhang, and J. Bao, Light-weight silver plating foam and carbon nanotube hybridized epoxy composite foams with exceptional conductivity and electromagnetic shielding property, *ACS Appl. Mater. Interfaces*, 8: 24131–24142, 2016. doi:10.1021/acsami.6b08325.

158. Q. Li, L. Chen, X. Li, J. Zhang, X. Zhang, K. Zheng et al., Effect of multi-walled carbon nanotubes on mechanical, thermal and electrical properties of phenolic foam via in-situ polymerization, *Compos. Part A Appl. Sci. Manuf.*, 82: 214–225, 2016. doi:10.1016/j.compositesa.2015.11.014.

159. J. Li, G. Zhang, Z. Ma, X. Fan, X. Fan, J. Qin et al., Morphologies and electromagnetic interference shielding performances of microcellular epoxy/multi-wall carbon nanotube nanocomposite foams, *Compos. Sci. Technol.*, 129: 70–78, 2016. doi:10.1016/j.compscitech.2016.04.003.

160. T. Kuang, L. Chang, F. Chen, Y. Sheng, D. Fu, and X. Peng, Facile preparation of lightweight high-strength biodegradable polymer/multi-walled carbon nanotubes nanocomposite foams for electromagnetic interference shielding, *Carbon*, 105: 305–313, 2016. doi:10.1016/j.carbon.2016.04.052.

9

Design for Improved Damage Resistance and Damage Tolerance of Polymer Matrix Composites

9.1 Introduction

The resistance offered by a material toward nucleation of any structural damage is known as damage resistance. In most of the cases, this is governed by the strength of the material. However, there may be several instances where the material contains some damage/crack during its fabrication, handling, or perhaps during its service. It is very important to evaluate the capability of a structural material to hold or to bear any structural defect/discontinuity during its service without or with very little change in its load-bearing capacity, which is known as "damage tolerance of the material." This is to ensure sufficient warning and, subsequently, enough time for the replacement of the component before catastrophic failure.

Delamination is one of the most life-limiting damage modes in laminated composite materials.

During processing, handling, or in-service conditions, delamination can occur, thus hampering the durability and structural integrity of the material. The reliability of laminated composites is harmfully affected by delamination introduction at the time of processing or the in-service environment. During processing, delamination may exist as voids, discontinuity in the material. External events like low-velocity impact or residual stresses generated from moisture, temperature, and so forth may result in delamination. Superior in-plane properties are exhibited by laminated composites, but out-of-plane stress fields can develop due to internal discontinuities even though the remote load applied externally is in-plane, as shown in Figure 9.1.

There are various ways to determine the damage resistance and damage tolerance of fiber-reinforced polymer (FRP) composites.

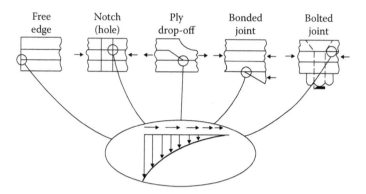

FIGURE 9.1
Various sources of out-of-plane stress development in composite structure. (From Sela, N. and Ishai, O., *Composites*, 20, 423–435, 1989.)

9.2 Methods to Determine Damage Tolerance

The intensive stress field ahead of the delamination/crack tip is responsible for its growth. Extension or propagation of this existing delamination may eventually arrive at a product with inferior mechanical strength. Delamination resistance has proven to be one of the major life-limiting criteria for laminated composite materials, because of which the application of composites was earlier limited to secondary structural parts of aerospace industries, where the loading conditions are well established and failure induced from loading is not life-threatening [2].

In addition to delamination matrix cracking, debonding at the matrix–fiber interface, fiber fracture, fiber pullout, and so on are some of the fracture modes in FRP composites [3]. Again, there are several mechanisms, which when combined, contribute toward the toughness of FRP composites. There may be a crack deflection, which can occur for a tilting or twisting movement around the fiber. A pair of new superficial areas is thus created, which apparently results in high fracture energy, as observed in the debonding mechanism. In a fiber-bridging mechanism, the unfractured, interfacial-slipped fibers establish a connection between both crack surfaces [4]. Environmental moisture plays an important role in determining the fracture behavior of FRP composites during their service period, as polymer matrix absorbs moisture. This becomes more important in the case of natural fiber-reinforced polymer composites [5].

Inherently, structural composite materials are subjected to three-dimensional (3D) loading. Thus, the existing delamination will be subjected to various modes of crack propagation: (1) crack opening or mode I, (2) forward shear or mode II, and (3) anti-plane shear or mode III, as shown in Figure 9.2. In practice, delamination is always of mixed mode type, as it is arrested between two

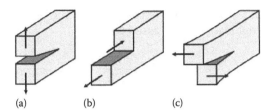

FIGURE 9.2
Modes of interlaminar crack propagation: (a) mode I opening mode, (b) mode II sliding shear mode, and (c) mode III tearing mode. (From Orifici, A. C. et al., *Mech. Compos. Mater.*, 43, 9–28, 2007.)

consecutive plies. Based upon the fundamentals of classical fracture mechanics, several tests have been standardized to study the resistance or tolerance power of the material under different fracture modes. The fundamental fracture mechanics concepts remain the same for all modes. However, the majority of the failure in the materials takes places in the mode I.

9.2.1 Mode I Fracture Test

The mode I fracture test has gained substantial attention, as this is directly related to the opening up of the crack. The most commonly used experimental technique to carry this out is the double cantilever beam (DCB) test, where the DCB specimen is loaded symmetrically under tension with a direction perpendicular to the plane of the crack. A typical DCB sample and testing technique (ASTM 5528 [7]) is shown in Figure 9.3 [8]. A nonadhesive Teflon film is placed in the midplane of the laminate during fabrication, which acts as delamination. The sample is then attached to two hinges, as shown in Figure 9.3. The hinges are then pulled away from each other and a load-displacement curve is thus generated.

After obtaining the load-displacement curve, the critical strain energy release rate (G_{IC}), which is a direct measure of fracture toughness, can directly

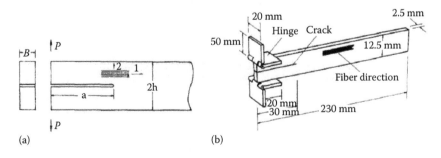

FIGURE 9.3
Mode I fracture testing: (a) a schematic of DCB specimen and (b) specimen dimensions. (From Ye, L., *Compos. Sci. Technol.*, 43, 49–54, 1992.)

be evaluated from the strain energy contained in the test specimen or work done by the external loads as expressed below [9].

$$G_{IC} = \frac{P_c^2}{2B} \frac{\partial C}{\partial a} \qquad (9.1)$$

where:
  P_c is the critical load applied
  B is the width of the beam
  $C\ (= \delta/P)$ is the compliance
  a is the crack length

The compliance is obtained from the slope of the loading/unloading line of the load-displacement curve.

9.2.2 Mode II Fracture Test

Interlaminar fracture toughness in the sliding shear mode is termed as a mode II fracture, which is commonly determined by an end-notched flexure (ENF) test. A 3-point bend specimen containing an embedded delamination at the midplane of the laminate (where interplaner shear stress is maximum) is termed as an "ENF specimen." A typical ENF sample with loading direction is shown in Figure 9.4.

The critical strain energy release rate for a mode II fracture (G_{IIC}) can be determined from the following expression [2]:

$$G_{IIC} = \frac{9a^2 P_c \delta}{2B\left(2L^2 + 3a^3\right)} \qquad (9.2)$$

where:
  P_c is the critical load applied
  δ is the displacement
  a is the crack length
  L and B are the length and width of the beam, respectively

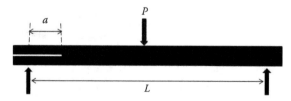

FIGURE 9.4
Mode II fracture toughness testing using an ENF sample.

9.2.3 Mode III Fracture Test

Typically, the mode III fracture is obtained by an edge crack torsion test. The test specimen and loading conditions are shown in Figure 9.5. The Irwin–Kies relation [10], shown below, calculates the mode III fracture toughness.

$$G_{IIIC} = \frac{mP_c^2}{2C(A-ma)^2} \tag{9.3}$$

where:
  C is the compliance

$$A = \frac{32\mu_{xy,0}h^3b}{3cd^2}$$

$$m = \frac{32\mu_{xy,0}h^3}{3cd^2}\left(1 - \frac{\mu_{xy,1}}{4\mu_{xy,0}}\right)$$

$\mu_{xy,0}$ and $\mu_{xy,1}$ represent cross-laminated timber torsional shear moduli of the uncracked and cracked parts of the specimen, respectively

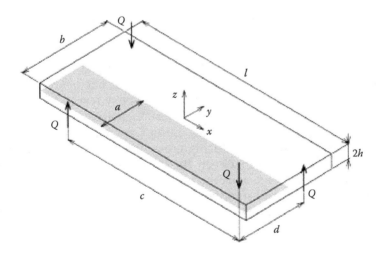

FIGURE 9.5
Mode III fracture toughness testing using edge crack torsion specimen. (From de Morais, A. B. et al., *Compos. Sci. Technol.*, 69, 670–676, 2009.)

9.2.4 Compression after Impact

Damages produced in FRP composites during processing, handling, or in-service conditions may have a pronounced detrimental effect on their structural integrity and durability. Damages may present in the structure in the form of discontinuity during processing, accidental dropping of heavy tools on the FRP composites, stress intensification at several designed holes, notches, and so on, or may be because of a low-velocity impact during service. Such damages may be invisible or barely visible, but can deteriorate the composite strength significantly, especially in compression. Various aircraft industries carry out tests to determine the compressive strength of the FRP composite after a low-velocity impact, termed as "compression after impact" testing, which is a measure of the damage-tolerating capacity of the composite [11]. After giving a low-velocity/energy impact, as per ASTM D5628-07, a compression test is carried out according to ASTM D7137 M-07, from which the compression after impact strength is determined using Equation 9.4 [12].

$$P_c = \frac{8\pi^2 E t^3 G_{IIc}}{9\left(1-v^2\right)} \tag{9.4}$$

9.3 Techniques for Improving Damage Tolerance

9.3.1 Toughening of Matrix

Fracture toughness of the epoxy resins is around 100 J/m² or less and therefore is applicable in aerospace industries. It has been shown that with an increase in the fracture toughness of epoxy resins, the interlaminar fracture toughness of the composite also increases. For an increase in the fracture toughness of the resins, various operations are performed, such as adding elastomers, reducing cross-link density, and increasing the resin chain flexibility between the cross-links. Adding rigid thermoplastic resins, such as polyetheretherketone, polyphenylene sulfide, and polyamide-imide and so on, that have fracture toughness of around 1,000 J/m² (i.e., 10 times higher than that of epoxy) are used [13] to achieve toughness.

9.3.2 Interleaving

Interleaving or adding a tough adhesive interlayer between laminate plies is one of the most promising method by which delamination resistance may be enhanced significantly without sacrificing the hot/wet performance of the composite material [14,15]. The restriction in the crack front-yield zone due to the formation of a narrow resin-rich region between rigid elastic plies, limits the translation of improved matrix toughness into improved composite toughness. About a 25-fold increase of matrix toughness can only translate to about four to eight times of toughness of a composite [16]. Interleaving allows more expansion of the crack-tip-yield zone between the composite plies [14]. However, the problem associated with this technique is the weight penalty to a structure because a tough layer of resin is introduced. Shear-yielding of the material around the crack tip can be attributed as the main mechanism responsible for toughness enhancement. A greater fracture energy due to the insertion of interleaving results due to the formation of a larger height of the crack-tip-yield zone.

9.3.3 Sequential Stacking

High interlaminar normal and shear stresses are created at the free edges of a laminate owing to a mismatch of Poisson's ratio and the coefficients of mutual influence between the adjacent layers. Therefore, generally, the ply sequence is altered to make a change in the state of stress from tensile form to compressive form, suppressing an opening mode delamination [13].

9.3.4 Interply Hybridization

This technique is also used to reduce the mismatch of Poisson's ratio and coefficients of mutual influence between the consecutive layers, thus eliminating the likelihood of interply edge disorders [13].

9.3.5 Through-the-Thickness Reinforcement

There are some well-explored strategies to enhance the delamination resistance and impact damage tolerance for textile laminates made using a dry fabric preform that contains the through-thickness reinforcement prior to resin infusion. These methods include, but are not limited to, 3D weaving, braiding, and stitching [17,18]. More specialist techniques include tufting, embroidery, and z-anchoring. However, none of these techniques are suitable for the through-thickness reinforcement of prepreg laminates. Z-pinning is the only technique which is capable of reinforcing prepreg laminates in the through-thickness direction. Z-pins [19] act as fine nails that lock the laminate plies together by a combination of friction and adhesion (Figure 9.6). Thin metal rods were used to reinforce the laminates by using a labor-intensive manual process, which is not practical for commercial production, initially, in the 1970s [20]. Tomashevkii et al. [21] developed an automated process for inserting thin wire fibers through laminates.

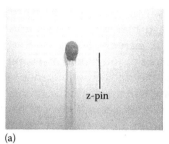
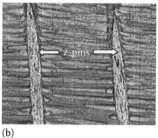

(a)   (b)

FIGURE 9.6
(a) Photograph showing the size of a typical z-pin and (b) z-pins inside a prepreg composite.
(From Mouritz, A. P., *Compos. Part A Appl. Sci. Manuf.*, 38, 2383–2397, 2007.)

9.3.6 Fiber Surface Modification

Most of the properties of FRP composites are limited by the interface. Fiber surface modification is one of the methods to control the interfacial properties. Various fiber surface treatment techniques, for example, reaction barrier coatings, have been developed to strengthen the interface. The choice of the technique adopted is decided depending on the compatibility and durability of the technique with the fiber type (composition and topology), matrix material. Coating is especially profitable when fibers like glass are exposed to a humid atmosphere where there is a thermodynamic tendency of the fiber to pick up water, resulting in a weak and porous surface. Silane coating on a glass fiber surface has been proven to be extremely beneficial for having good interfacial bonding. One end of the silane coupling agent adheres with the fiber and at the other end forms a strong linkage with the polymer matrix. Interfacial properties like inter-laminar shear strength (ILSS) and fracture toughness (K_{IC}) are improved using a silane coupling agent on a glass fiber surface [22]. It has further been observed that a higher interfacial shear stress is transmitted in the case of silane coating [23]. An electrochemical alternating anode and cathode treatment [24] of carbon fibers in various electrolytes like H_2SO_4 are proven to increase the surface grooves on carbon fibers, which act as a physical anchor hold between the fiber and the matrix, resulting in a superior bonding at the interface.

9.3.7 Fiber Architecture

The widespread adoption and acceptability of 2D laminated composites in some critical structures of automobiles and aircraft also have been limited by their low through-thickness mechanical properties and inferior impact damage resistance when compared against their traditional metallic counterparts, such as aluminum alloys and steel. Another important advantage of 3D woven composites is their low-velocity impact damage tolerance [25,26] and high ballistic impact damage resistance [27], which is the major problem, associated with 2D laminates in high performance structures. Chou et al. [25] suggested that the impact energy required to initiate damage in 3D woven carbon–bismaleimide composites is about 60% higher than in a 2D carbon–bismaleimide laminate. 3D composites offer high impact damage resistance and simultaneously they do not undergo reduction in their in-plane mechanical properties as compared to their 2D laminate counterparts [28,29]. The high damage tolerance of 3D composites results due to the through-thickness binder yarns. These yarns are able to arrest or retard the growth of delamination cracks generated under an impact loading. The binder yarns sometimes also results in increased tensile strain-to-failure values [30]. Their mode I interlaminar fracture toughness values have been reported as 6–20 times higher than the unidirectional carbon fiber-reinforced epoxy laminates [31]. Despite having these superior properties and potential benefits, the 3D woven composites have failed to find many commercial applications.

9.3.8 Nanocomposite

As discussed in Chapter 8, incorporation of the carbon nanotube kind of nanofillers in polymer composites not only increase its strength, but also its damage tolerance by several toughening mechanisms like crack bridging, crack pinning, and so on.

References

1. N. Sela and O. Ishai, Interlaminar fracture toughness and toughening of laminated composite materials: A review, *Composites*, 20: 423–435, 1989. doi:10.1016/0010-4361(89)90211-5.
2. J.-K. Kim and Y.-W. Mai, *Engineered Interfaces in Fiber Reinforced Composites*. Amsterdam, the Netherlands: Elsevier, 1998.
3. C. K. H. Dharan, Fracture mechanics of composite materials, *J. Eng. Mater. Technol.*, 100, 233–247, 1978.
4. R. V. Silva, D. Spinelli, W. W. Bose Filho, S. Claro Neto, G. O. Chierice, and J. R. Tarpani, Fracture toughness of natural fibers/castor oil polyurethane composites, *Compos. Sci. Technol.*, 66: 1328–1335, 2006. doi:10.1016/j.compscitech.2005.10.012.
5. G. J. Dvorak, Composite materials: Inelastic behavior, damage, fatigue and fracture, *Int. J. Solids Struct.*, 37: 155–170, 2000. doi:10.1016/S0020-7683(99)00085-2.
6. A. C. Orifici, R. S. Thomson, R. Degenhardt, C. Bisagni, and J. Bayandor, Development of a finite-element analysis methodology for the propagation of delaminations in composite structures, *Mech. Compos. Mater.*, 43: 9–28, 2007. doi:10.1007/s11029-007-0002-6.
7. ASTM, *Standard Test Method for Mode I Interlaminar Fracture Toughness of Unidirectional Fiber-Reinforced Polymer Matrix Composites*. Philadelphia, PA: ASTM, 1994.

8. L. Ye, Evaluation of Mode-I interlaminar fracture toughness for fiber-reinforced composite materials, *Compos. Sci. Technol.*, 43: 49–54, 1992. doi:10.1016/0266-3538(92)90132-M.

9. G. Caprino, The use of thin DCB specimens for measuring mode I interlaminar fracture toughness of composite materials, *Compos. Sci. Technol.*, 39: 147–158, 1990. doi:10.1016/0266-3538(90)90052-7.

10. A. B. de Morais, A. B. Pereira, M. F. S. F. de Moura, and A. G. Magalhães, Mode III interlaminar fracture of carbon/epoxy laminates using the edge crack torsion (ECT) test, *Compos. Sci. Technol.*, 69: 670–676, 2009. doi:10.1016/j.compscitech.2008.12.019.

11. Y. Tang, L. Ye, Z. Zhang, and K. Friedrich, Interlaminar fracture toughness and CAI strength of fibre-reinforced composites with nanoparticles: A review. *Compos. Sci. Technol.*, 86: 26–37, 2013. doi:10.1016/j.compscitech.2013.06.021.

12. V. Kostopoulos, A. Baltopoulos, P. Karapappas, A. Vavouliotis, and A. Paipetis, Impact and after-impact properties of carbon fibre reinforced composites enhanced with multi-wall carbon nanotubes. *Compos. Sci. Technol.*, 70: 553–563, 2010. doi:10.1016/j.compscitech.2009.11.023.

13. P. K. Mallick, *Fiber-Reinforced Composites: Materials, Manufacturing, and Design*. Boca Raton, FL: CRC Press, 2007.

14. P. A. Lagace and M. J. Kraft, Impact response of graphite/epoxy fabric structures, NASA. Langley Research Center, Eighth DOD(NASA)FAA Conference on Fibrous Composites in Structural Design, Part 2, 559–571, 1990.

15. W. L. Bradley and R. N. Cohen, Matrix deformation and fracture in graphite-reinforced epoxies, *In Delamination and debonding of materials*. ASTM International, 1985. doi:10.1520/STP36316S.

16. D. L. Hunston, Composite interlaminar fracture: Effect of matrix fracture energy. *J. Compos. Technol. Res.*, 6: 176–180, 1984. doi:10.1520/CTR10842J.

17. L. Tong, A. P. Mouritz, and M. K. Bannister, *3D Fibre Reinforced Polymer Composites*. Amsterdam, the Netherlands: Elsevier, 2002.

18. K. Dransfield, C. Baillie, and Y.-W. Mai, Improving the delamination resistance of CFRP by stitching: A review, *Compos. Sci. Technol.*, 50: 305–317, 1994. doi:10.1016/0266-3538(94)90019-1.

19. A. P. Mouritz, Review of z-pinned composite laminates, *Compos. Part A Appl. Sci. Manuf.*, 38: 2383–2397, 2007. doi:10.1016/j.compositesa.2007.08.016.

20. S. L. Huang, Cross reinforcement in a Gr/Ep laminate, *Paper Presented at American Society of Mechanical Engineering*, San-Francisco, CA, 1978.

21. V. T. Tomashevskii, S. Y. Sitnikov, V. N. Shalygin, and V. S. Yakovlev. A method of calculating technological regimes of transversal reinforcement of composites with short-fiber microparticles, *Mech. Compos. Mater.*, 25: 400–406, 1989. doi:10.1007/BF00614810.

22. S.-J. Park and J.-S. Jin, Effect of silane coupling agent on interphase and performance of glass fibers/unsaturated polyester composites, *J. Colloid Interface Sci.*, 242: 174–179, 2001. doi:10.1006/jcis.2001.7788.

23. A. T. Dibenedetto and P. J. Lex, Evaluation of surface treatments for glass fibers in composite materials, *Polym. Eng. Sci.*, 29: 543–555, 1989. doi:10.1002/pen.760290809.

24. Y. Ma, J. Wang, and X. Cai, The effect of electrolyte on surface composite and microstructure of carbon fiber by electrochemical treatment, *Int. J. Electrochem. Sci.*, 8: 2806–2815, 2013.

25. S. Chou, H.-C. Chen, and C.-C. Wu, BMI resin composites reinforced with 3D carbon-fibre fabrics, *Compos. Sci. Technol.*, 43: 117–128, 1992. doi:10.1016/0266-3538(92)90002-K.
26. D. P. C. Aiman, M. F. Yahya, and J. Salleh, Impact properties of 2D and 3D woven composites: A review, *AIP Conf. Proc.*, 1774: 020002, 2016. doi:10.1063/1.4965050.
27. W. E. Lundblad, C. Dixon, and H. C. Ohler, Ballistic resistant article comprising a three dimensional interlocking woven fabric. US5456974 A, 1995.
28. L. Dickinson, M. H. Mohammed, and E. Klang, Impact resistance and compressional properties of three-dimensional woven carbon/epoxy composites, in *Developments in the Science and Technology of Composite Materials*. Dordrecht, the Netherlands: Springer, pp. 659–664, 1990. doi:10.1007/978-94-009-0787-4_91.
29. G. L. Farley, B. T. Smith, and J. Maiden, Compression response of thick layer composite laminates with through-the-thickness reinforcement, *J. Reinf. Plast. Compos.*, 11: 787–810, 1992. doi:10.1177/073168449201100705.
30. B. N. Cox, M. S. Dadkhah, and W. L. Morris, On the tensile failure of 3D woven composites, *Compos. Part A Appl. Sci. Manuf.*, 27: 447–458, 1996. doi:10.1016/1359-835X(95)00053-5.
31. A. P. Mouritz, C. Baini, and I. Herszberg, Mode I interlaminar fracture toughness properties of advanced textile fibreglass composites, *Compos. Part A Appl. Sci. Manuf.*, 30: 859–870, 1999. doi:10.1016/S1359-835X(98)00197-3.

Index

Note: Page numbers followed by f and t refer to figures and tables respectively

Milton Keynes UK
Ingram Content Group UK Ltd.
UKHW040059071024
449327UK00019B/668